MEMOIRS
of the
American Mathematical Society

ISSN 0065-9266 (print)
ISSN 1947-6221 (online)

Number 1371

AF739135

Cancellation for Surfaces Revisited

H. Flenner
S. Kaliman
M. Zaidenberg

July 2022 • Volume 278 • Number 1371 (sixth of 6 numbers)

AMERICAN MATHEMATICAL SOCIETY

Library of Congress Cataloging-in-Publication Data

Cataloging-in-Publication Data has been applied for by the AMS.
See http://www.loc.gov/publish/cip/.
DOI: https://doi.org/10.1090/memo/1371

Memoirs of the American Mathematical Society

This journal is devoted entirely to research in pure and applied mathematics.

Subscription information. Beginning with the January 2010 issue, *Memoirs* is accessible from `www.ams.org/journals`. The 2022 subscription begins with volume 275 and consists of six mailings, each containing one or more numbers. Subscription prices for 2022 are as follows: for paper delivery, US$1085 list, US$868 institutional member; for electronic delivery, US$955 list, US$764 institutional member. Upon request, subscribers to paper delivery of this journal are also entitled to receive electronic delivery. If ordering the paper version, add US$22 for delivery within the United States; US$85 for outside the United States. Subscription renewals are subject to late fees. See `www.ams.org/help-faq` for more journal subscription information. Each number may be ordered separately; *please specify number* when ordering an individual number.

Back number information. For back issues see `www.ams.org/backvols`.

Subscriptions and orders should be addressed to the American Mathematical Society, P. O. Box 845904, Boston, MA 02284-5904 USA. *All orders must be accompanied by payment.* Other correspondence should be addressed to 201 Charles Street, Providence, RI 02904-2213 USA.

Memoirs of the American Mathematical Society (ISSN 0065-9266 (print); 1947-6221 (online)) is published bimonthly (each volume consisting usually of more than one number) by the American Mathematical Society at 201 Charles Street, Providence, RI 02904-2213 USA. Periodicals postage paid at Providence, RI. Postmaster: Send address changes to Memoirs, American Mathematical Society, 201 Charles Street, Providence, RI 02904-2213 USA.

This publication is indexed in *Mathematical Reviews®*, *Zentralblatt MATH*, *Science Citation Index®*, *Science Citation Index™-Expanded*, *ISI Alerting Services℠*, *SciSearch®*, *Research Alert®*, *CompuMath Citation Index®*, *Current Contents®/Physical, Chemical & Earth Sciences.*
This publication is archived in *Portico* and *CLOCKSS*.
Printed in the United States of America.

∞ The paper used in this book is acid-free and falls within the guidelines
established to ensure permanence and durability.
Visit the AMS home page at `https://www.ams.org/`

10 9 8 7 6 5 4 3 2 1 27 26 25 24 23 22

Contents

Introduction 1

Chapter 1. Generalities 5
1.1. Cancellation and the Makar-Limanov invariant 5
1.2. Non-cancellation and Gizatullin surfaces 6
1.3. The Danielewski–Fieseler construction 6
1.4. Affine modifications 7

Chapter 2. $\mathbb{A}^1$-fibered surfaces via affine modifications 9
2.1. Covering trick and GDF surfaces 9
2.2. Pseudominimal completion and extended divisor 11
2.3. Blowup construction 12
2.4. GDF surfaces via affine modifications 16

Chapter 3. Vector fields and natural coordinates 21
3.1. Locally nilpotent vertical vector fields 21
3.2. Standard affine charts 21
3.3. Natural coordinates 22
3.4. Special μ_d-quasi-invariants 23
3.5. Examples of GDF surfaces of Danielewski type 24

Chapter 4. Relative flexibility 27
4.1. Definitions and the main theorem 27
4.2. Transitive group actions on Veronese cones 28
4.3. Relatively transitive group actions on cylinders 30
4.4. A relative Abhyankar-Moh-Suzuki Theorem 32

Chapter 5. Rigidity of cylinders upon deformation of surfaces 35
5.1. Equivariant Asanuma modification 35
5.2. Rigidity of cylinders under deformations of GDF surfaces 37
5.3. Rigidity of cylinders under deformations of $\mathbb{A}^1$-fibered surfaces 39
5.4. Rigidity of line bundles over affine surfaces 40

Chapter 6. Basic examples of Zariski factors 47
6.1. Line bundles over affine curves 47
6.2. Parabolic $\mathbb{G}_m$-surfaces: an overview 48
6.3. Parabolic $\mathbb{G}_m$-surfaces as Zariski factors 51

Chapter 7. Zariski 1-factors 59
7.1. Stretching and rigidity of cylinders 59
7.2. Non-cancellation for GDF surfaces 67

7.3. Extended graphs of Gizatullin surfaces 69
7.4. Zariski 1-factors and affine $\mathbb{A}^1$-fibered surfaces 71

Chapter 8. Classical examples 73

Chapter 9. GDF surfaces with isomorphic cylinders 77
9.1. Preliminaries 77
9.2. Classification of GDF cylinders up to B-isomorphism 78
9.3. GDF surfaces whose fiber trees are bushes 79
9.4. Spring bushes versus bushes 82
9.5. Cylinders over Danielewski-Fieseler surfaces 89
9.6. Proof of the main theorem 90

Chapter 10. On moduli spaces of GDF surfaces 97
10.1. Coarse moduli spaces of GDF surfaces 97
10.2. The automorphism group of a GDF surface 98
10.3. Configuration spaces and configuration invariants 99
10.4. Versal deformation families of trivializing sequences 102
10.5. Proof of Theorem 10.1.3 106
Acknowledgments 107

Bibliography 109

Abstract

The celebrated Zariski Cancellation Problem asks as to when the existence of an isomorphism $X \times \mathbb{A}^n \cong X' \times \mathbb{A}^n$ for (affine) algebraic varieties X and X' implies that $X \cong X'$. In this paper we provide a criterion for cancellation by the affine line (that is, $n = 1$) in the case where X is a normal affine surface admitting an $\mathbb{A}^1$-fibration $X \to B$ with no multiple fiber over a smooth affine curve B. For two such surfaces $X \to B$ and $X' \to B$ we give a criterion as to when the cylinders $X \times \mathbb{A}^1$ and $X' \times \mathbb{A}^1$ are isomorphic over B. The latter criterion is expressed in terms of linear equivalence of certain divisors on the Danielewski-Fieseler quotient of X over B. It occurs that for a smooth $\mathbb{A}^1$-fibered surface $X \to B$ the cancellation by the affine line holds if and only if $X \to B$ is a line bundle, and, for a normal such X, if and only if $X \to B$ is a cyclic quotient of a line bundle (an orbifold line bundle). If X does not admit any $\mathbb{A}^1$-fibration over an affine base then the cancellation by the affine line is known to hold for X by a result of Bandman and Makar-Limanov.

If the cancellation does not hold then X deforms in a non-isotrivial family of $\mathbb{A}^1$-fibered surfaces $X_\lambda \to B$ with cylinders $X_\lambda \times \mathbb{A}^1$ isomorphic over B. We construct such versal deformation families and their coarse moduli spaces provided B does not admit nonconstant invertible functions. Each of these coarse moduli spaces has infinite number of irreducible components of growing dimensions; each component is an affine variety with quotient singularities. Finally, we analyze from our viewpoint the examples of non-cancellation constructed by Danielewski, tom Dieck, Wilkens, Masuda and Miyanishi, e.a.

Received by the editor April 7, 2018, and, in revised form, April 16, 2019.

Article electronically published on June 6, 2022.

DOI: https://doi.org/10.1090/memo/1371

2020 *Mathematics Subject Classification.* Primary 14R10; Secondary 14D22.

Key words and phrases. Cancellation, affine surface, group action, one-parameter subgroup, transitivity.

Introduction

Let X and Y be algebraic varieties over a field $\Bbbk$. The celebrated Zariski Cancellation Problem, in its most general form, asks under which circumstances the existence of a biregular (resp., birational) isomorphism $X \times \mathbb{A}^n \cong Y \times \mathbb{A}^n$ implies that $X \cong Y$, where $\mathbb{A}^n$ stands for the affine n-space over $\Bbbk$. In this paper we are interested in the biregular cancellation problem, hence the symbol '$\cong$' stands for a biregular isomorphism. We say that X is a *Zariski factor* if X is *cancellative*, that is, whenever Y is an algebraic variety, $X \times \mathbb{A}^n \cong Y \times \mathbb{A}^n$ implies $X \cong Y$ for any $n \in \mathbb{N}$. We say that X is a *strong Zariski factor* if any isomorphism $\Phi\colon X \times \mathbb{A}^n \to Y \times \mathbb{A}^n$ fits in a commutative diagram

$$\begin{array}{ccc} X \times \mathbb{A}^n & \xrightarrow{\Phi} & Y \times \mathbb{A}^n \\ \downarrow & & \downarrow \\ X & \xrightarrow[\varphi]{\cong} & Y \end{array}$$

where the vertical arrows are the canonical projections. This property is usually called a *strong cancellation*. We say that X is a *Zariski* 1*-factor* if $X \times \mathbb{A}^1 \cong Y \times \mathbb{A}^1$ implies that $X \cong Y$, and a *strong Zariski* 1*-factor* if the strong cancellation holds for X with $n = 1$. The latter implies that the cylinder structure on $X \times \mathbb{A}^1$ is unique, see [**50**, Thm. 2.18].

By a theorem of Abhyankar, Heinzer and Eakin ([**1**, Thm. 6.5]) any affine curve C is a Zariski factor, and if $C \ncong \mathbb{A}^1$ then C is a strong Zariski factor. More generally, by the Iitaka-Fujita Theorem ([**44**]) any algebraic variety of non-negative log-Kodaira dimension is a strong Zariski factor. Due to a theorem by Bandman and Makar-Limanov ([**9**, Lem. 2][1]) the following holds.

THEOREM 0.0.1 (Bandman and Makar-Limanov). *The affine varieties which do not admit any effective $\mathbb{G}_a$-action are strong Zariski* 1*-factors.*

There are examples of smooth rational affine surfaces of negative log-Kodaira dimension which are $\mathbb{A}^1$-fibered over $\mathbb{P}^1$ and do not admit any effective $\mathbb{G}_a$-action, and so, are strong Zariski 1-factors, see [**9**, Ex. 3], [**41**, 3.7]. Some of these affine surfaces are not Zariski 2-factors, see [**23**, **24**].

In this paper we concentrate on the Zariski Cancellation Problem for normal affine surfaces over an algebraically closed field $\Bbbk$ of characteristic zero. From Theorem 0.0.1 one can deduce the following criteria.

COROLLARY 0.0.2. *A normal affine surface X is a strong Zariski* 1*-factor if and only if it does not admit any effective $\mathbb{G}_a$-action, if and only if it is not fibered over a smooth affine curve C with general fibers isomorphic to the affine line $\mathbb{A}^1$.*

[1] Cf. [**18**]; see [**13**, Thm. 3.1] for the positive characteristic case.

See, e.g., [**50**, Thm. 2.18] for the first part and [**29**, Lem. 1.6] for the second.

Recall (see e.g., [**28**]) that a *parabolic* $\mathbb{G}_m$*-surface* is a normal affine surface X equipped with an $\mathbb{A}^1$-fibration $\pi: X \to C$ over a smooth affine curve C and with an effective $\mathbb{G}_m$-action along the fibers of π. Any fiber of π on such a surface X is isomorphic to $\mathbb{A}^1$. There is exactly one singular point of X in each multiple fiber of π and no further singularities. Any singular point $x \in X$ is a cyclic quotient singularity. If a parabolic $\mathbb{G}_m$-surface $X \to C$ is smooth then this is a line bundle over C. Any parabolic $\mathbb{G}_m$-surface admits an effective $\mathbb{G}_a$-action along the fibers of π ([**29**, Thm. 3.12]).

By the celebrated Miyanishi-Sugie-Fujita Theorem ([**35, 57**]; see also [**56**, Ch. 3, Thm. 2.3.1]) the affine plane $\mathbb{A}^2$ is a Zariski factor. An analogous result holds for the parabolic $\mathbb{G}_m$-surfaces. Moreover, the following criterion holds.

THEOREM 0.0.3. *For a normal affine surface X equipped with an $\mathbb{A}^1$-fibration $X \to C$ over a smooth affine curve C the following conditions are equivalent:*

(i) *X is a Zariski factor;*
(ii) *X is a Zariski 1-factor;*
(iii) *X is a parabolic $\mathbb{G}_m$-surface.*

The implication (i)⇒(ii) is immediate; see Theorem 7.4.1 for (ii)⇒(iii) and Theorem 6.3.1 for (iii)⇒(i).

From Theorems 0.0.1 and 0.0.3 one can deduce the following characterization.

COROLLARY 0.0.4. *A normal affine surface X is a Zariski 1-factor if and only if either X does not admit any effective $\mathbb{G}_a$-action, or X is a parabolic $\mathbb{G}_m$-surface.*

The Danielewski surfaces

$$X_m = \{z^m t - u^2 - 1 = 0\} \subset \mathbb{A}^3, \qquad m \in \mathbb{N},$$

are examples of non-Zariski 1-factors ([**17, 26**]). Being pairwise non-homeomorphic ([**26**]) these surfaces have isomorphic cylinders: $X_m \times \mathbb{A}^1 \cong X_{m'} \times \mathbb{A}^1 \ \forall m, m' \in \mathbb{N}$. For non-Zariski factors we consider the following problem.

0.0.5. **Problem.** *Given an affine algebraic variety X, describe the moduli space $\mathcal{C}_m(X)$ of isomorphism classes of the affine algebraic varieties Y such that $X \times \mathbb{A}^m \cong Y \times \mathbb{A}^m$. Study the behavior of $\mathcal{C}_m(X)$ upon deformation of X.*

Note that X is a Zariski 1-factor if and only if $\mathcal{C}_1(X) = \{X\}$. There is no example of an affine non-Zariski 1-factor X for which the moduli space $\mathcal{C}_1(X)$ were known. For the first Danielewski surface X_1 the moduli space $\mathcal{C}_1(X_1)$ has infinite number of irreducible components. In [**69**] and [**54**, Thm. 2.8] this sequence is extended to a family of surfaces in $\mathbb{A}^3$ with similar properties. These examples show that $\mathcal{C}_1(X_1)$ possesses an infinite number of components which are infinite dimensional ind-varieties.

We show that in a majority of cases a normal affine surface $\mathbb{A}^1$-fibered over an affine curve deforms in a large family of such surfaces with isomorphic cylinders; see Theorems 5.2.2 and 5.3.1. Moreover, the deformation space contains infinitely many connected components of growing dimensions.

Let $\pi: X \to B$ be an $\mathbb{A}^1$-fibered surface over a smooth affine curve B. We call such a surface a *generalized Danielewski-Fieseler surface*, or a *GDF surface* for short, provided π has only reduced fibers. To a GDF surface $\pi: X \to B$ one associates a non-separated one-dimensional scheme $\mathrm{DF}(\pi)$ called the *Danielewski-Fieseler quotient* along with a surjective morphism $\mathrm{DF}(\pi) \to B$ and an anti-effective divisor $\mathrm{tp.div}(\pi)$ on $\mathrm{DF}(\pi)$ called the *type divisor* (see Definitions 7.1.8 and 7.1.9). In Section 9 we prove the following theorem.

THEOREM 0.0.6. *Let $\pi: X \to B$ and $\pi': X' \to B$ be GDF surfaces. Then the cylinders $X \times \mathbb{A}^1$ and $X' \times \mathbb{A}^1$ are isomorphic over B if and only if there exists an isomorphism $\tau: \mathrm{DF}(\pi) \xrightarrow{\cong_B} \mathrm{DF}(\pi')$ defined over B such that the divisors $\mathrm{tp.div}(\pi)$ and $\tau^*(\mathrm{tp.div}(\pi'))$ on $\mathrm{DF}(\pi)$ are linearly equivalent.*

The next corollary follows immediately by using a suitable base change.

COROLLARY 0.0.7. *An isomorphism $\varphi: X \times \mathbb{A}^1 \xrightarrow{\cong} X' \times \mathbb{A}^1$ which sends the fibers of $X \times \mathbb{A}^1 \to B$ to fibers of $X' \times \mathbb{A}^1 \to B$ does exist if and only if there exists an isomorphism $\tau: \mathrm{DF}(\pi) \xrightarrow{\cong} \mathrm{DF}(\pi')$ such that $\mathrm{tp.div}(\pi) \sim \tau^*(\mathrm{tp.div}(\pi'))$.*

REMARKS 0.0.8. 1. Notice that if $B \ncong \mathbb{A}^1$ then any isomorphism of cylinders $X \times \mathbb{A}^1 \xrightarrow{\cong} X' \times \mathbb{A}^1$ sends the fibers to fibers inducing an automorphism of B, cf., e.g., Lemma 6.3.4.

2. It is worth to mention also the following facts. Consider a pair $(\breve{B} \to B, \breve{D})$ where $\breve{D}$ is an anti-effective divisor on a one-dimensional (non-separated, in general) scheme $\breve{B}$ equipped with a birational surjective morphism $\breve{B} \to B$. Then there exists a GDF surface $\pi: X \to B$ such that $\mathrm{DF}(\pi) =_B \breve{B}$ and $\mathrm{tp.div}(\pi) = \breve{D}$. Given a pair $(\breve{B} \to B, \breve{D})$ the corresponding GDF surfaces X can vary in non-isotrivial families. However, due to Theorem 0.0.6 the cylinders over these surfaces are all isomorphic over B. Moreover, up to an isomorphism over B these cylinders depend only on the class of $\breve{D}$ in the Picard group $\mathrm{Pic}(\breve{B})$. The variation of $\breve{D}$ within its class adds, in general, extra discrete parameters to the isomorphism type of the corresponding GDF surface X, see Lemma 7.1.18 and Corollary 7.1.19.

Using Theorem 0.0.6 we provide in Section 9.2 a new proof of a result of Bandman and Makar-Limanov [**9**] which gives a sufficient condition for almost flexibility of the cylinder over an $\mathbb{A}^1$-fibered surface, see Theorem 9.2.3.

In the concluding Section 10 we construct a coarse moduli space of marked GDF surfaces with a given base B and a given graph divisor (see Definition 2.3.11), provided B does not admit nonconstant invertible functions, see Theorem 10.1.3. The cylinders over these surfaces are all isomorphic over B. A simple Example 10.5.3 shows that for B which does not satisfy the above assumption, the coarse moduli space of such surfaces does not exist, in general. The irreducible components of the moduli space of GDF surfaces with a given cylinder have unbounded dimensions. This resolves the first part of Problem 0.0.5; notice that the "isomorphism of cylinders over B" can be replaced by the "isomorphism" if $B \ncong \mathbb{A}^1$.

The proofs of the main results exploit the affine modifications ([**47**]), in particular, the Asanuma modification ([**6**]), and the flexibility techniques of [**2**], in particular, the interpolation by automorphisms. As an illustration, in Section 3.5 we analyze from our viewpoint the examples of non-cancellation due to Danielewski ([**17**]), Fieseler ([**26**]), Wilkens ([**69**]), tom Dieck ([**68**]), Miyanishi–Masuda ([**54**]), and the examples of Danielewski-Fieseler surfaces due to Dubouloz and Poloni ([**25**], [**61**]).

Being rather long, the proofs of our main results contain many constructive elements. Indeed, in Section 5 (see Theorem 5.2.2) we show that, up to isomorphism of cylinders, the continuous parameters of GDF surfaces can be ignored. Moreover, up to isomorphism of cylinders, the non-essential discrete parameters can also be ignored in a sense, see Chapter 9. It follows that, for a fixed base curve and fixed positions of degenerate fibers, the cylinders of our GDF surfaces form, up to isomorphism over the base, a sequence of affine threefolds which can be given by simple and canonical equations; see (9.4). Once their equations are written under the canonical form, one has a constructive way to pass from one cylinder to an isomorphic one, see Proposition 9.6.5.

REMARK 0.0.9. The results of the paper were reported by the third author on the conference "Complex analyses and dynamical systems - VII" (Nahariya, Israel, May 10–15, 2015), on a seminar at the Bar Ilan University (Ramat Gan, Israel, May 24, 2015), and in the lecture course "Affine algebraic surfaces and the Zariski cancellation problem" at the University of Rome Tor Vergata (September–November, 2015; see the program in [**71**]). When this paper was written the third author assisted at the lecture course by Adrien Dubouloz on the cancellation problem for affine surfaces in the 39th Autumn School in Algebraic Geometry (Łukęcin, Poland, September 19–24, 2016). In this course Adrien Dubouloz advertised a result on non-cancellation for smooth $\mathbb{A}^1$-fibered affine surfaces similar to our one (see, in particular, Theorem 1.3.1 below and Theorem 0.0.3 in the case of smooth surfaces), and indicated nice ideas of proofs done by completely different methods. He also posed the question whether the non-degenerate affine toric surfaces are Zariski 1-factors. This had been answered affirmatively by our Theorem 0.0.3.

CHAPTER 1

Generalities

1.1. Cancellation and the Makar-Limanov invariant

The special automorphism group $\mathrm{SAut}\,X$ of an affine variety X is the subgroup of the group $\mathrm{Aut}\,X$ generated by all its $\mathbb{G}_a$-subgroups ([**2**]). The *Makar-Limanov invariant* $\mathrm{ML}(X)$ is the subring of invariants of the action of $\mathrm{SAut}\,X$ on $\mathcal{O}_X(X)$. The $\mathrm{SAut}\,X$-orbits are locally closed in X ([**2**]). The *complexity* κ of the action of $\mathrm{SAut}\,X$ on X is the codimension of its general orbit, or, which is the same, the transcendence degree of the ring $\mathrm{ML}(X)$ ([**2**]). We design this integer κ as the *Makar-Limanov complexity* of X, and we say that X belongs to the class (ML_κ).

By the Miyanishi-Sugie Theorem ([**57**], [**56**, Ch. 2, Thm. 2.1.1, Ch. 3, Lem. 1.3.1 and Thm. 1.3.2]) a normal affine surface X with $\bar{k}(X) = -\infty$ contains a cylinder, that is, a principal Zariski open subset U of the form $U \cong C \times \mathbb{A}^1$ where C is a smooth affine curve. It possesses as well an $\mathbb{A}^1$-fibration $\mu\colon X \to B$ over a smooth curve B which extends the first projection $U \to C$ of the cylinder. If B is affine then X admits an effective action of the additive group $\mathbb{G}_a = \mathbb{G}_a(\mathbb{k})$ along the rulings of μ.

Conversely, suppose that there is an effective $\mathbb{G}_a$-action on X. Then the algebra of invariants $\mathcal{O}_X(X)^{\mathbb{G}_a}$ is finitely generated and normal ([**26**, Lem. 1.1]). Hence $B = \mathrm{Spec}\,\mathcal{O}_X(X)^{\mathbb{G}_a}$ is a smooth affine curve and the morphism $\mu\colon X \to B$ induced by the inclusion $\mathcal{O}_X(X)^{\mathbb{G}_a} \hookrightarrow \mathcal{O}_X(X)$ defines an $\mathbb{A}^1$-fibration (an affine ruling) on X. Such an $\mathbb{A}^1$-fibration is trivial over a Zariski open subset of B. It extends the first projection of a principal cylinder on X. If an $\mathbb{A}^1$-fibration on a surface X over an affine base is unique (non-unique, respectively) then X is of class (ML_1) (of class (ML_0), respectively). It is of class (ML_2) if X does not admit any $\mathbb{A}^1$-fibration over an affine curve. In the latter case X still could admit an $\mathbb{A}^1$-fibration over a projective curve. It does admit such a fibration if and only if $\bar{k}(X) = -\infty$.

The cancellation problem is closely related to the problem on stability of the Makar-Limanov invariant upon passing to a cylinder. The latter is discussed, e.g., in [**7**]–[**9**] and [**12**]-[**14**]. Suppose, for instance, that $\mathrm{ML}(X) = \mathcal{O}_X(X)$. Then by [**13**, Thm. 3.1] (cf. also [**18**]), $\mathrm{ML}(X \times \mathbb{A}^1) = \mathcal{O}_X(X)$. This means that the cylinder structure on $X \times \mathbb{A}^1$ is unique. Hence *an affine variety X which does not admit any effective $\mathbb{G}_a$-action is a Zariski 1-factor.* In particular, any smooth, affine surface of class (ML_2) is a Zariski 1-factor. Therefore, in the future we restrict to surfaces of classes (ML_0) and (ML_1), that is, to $\mathbb{G}_a$-surfaces.

In the Danielewski example, $X_1 \in (\mathrm{ML}_0)$ whereas $X_m \in (\mathrm{ML}_1)$ for $m \geq 2$. Thus, the Makar-Limanov complexity is not an invariant of cancellation (see also [**22**] for an example of the Koras-Russell cubic threefold). By contrast, the Euler characteristic, the Picard number (for a rational variety), the log-plurigenera, and the log-irregularity are cancellation invariants, see, e.g., Iitaka's Lemma in [**56**, Ch. 2, Lem. 1.15.1] and [**35**, (9.9)].

1.2. Non-cancellation and Gizatullin surfaces

Let X be a smooth affine surface. Recall ([**39**]) that $\mathrm{SAut}\,X$ acts on X with an open orbit if and only if $X \in \mathrm{ML}_0$. In the latter case X is a *Gizatullin surface*, i.e., a normal affine surface completable by a chain of smooth rational curves and different from $\mathbb{A}^1 \times (\mathbb{A}^1 \setminus \{0\})$. Furthermore, the group $\mathrm{SAut}(X \times \mathbb{A}^1)$ also acts with an open orbit on the cylinder $X \times \mathbb{A}^1$. Thus, the Makar-Limanov invariant $\mathrm{ML}(X \times \mathbb{A}^1)$ is trivial: $\mathrm{ML}(X \times \mathbb{A}^1) = \mathrm{ML}(X) = \Bbbk$.

The following conjecture is inspired by [**9**, §4, Thm. 1] and the unpublished notes [**10**] kindly offered to one of us by the authors. We use the notation of Problem 0.0.5.

CONJECTURE 1.2.1. *Let X be a normal affine surface such that the group $\mathrm{SAut}(X \times \mathbb{A}^1)$ acts with an open orbit in $X \times \mathbb{A}^1$. Then $\mathcal{C}_1(X)$ contains (the class of) a Gizatullin surface.*

Due to [**9**, Thm. 1] (see also an alternative proof in Part II) this conjecture is true for the Danielewski-Fieseler surfaces, that is, for the $\mathbb{A}^1$-fibered surfaces $\pi: X \to \mathbb{A}^1$ with a unique degenerated fiber, provided this fiber is reduced.

1.3. The Danielewski–Fieseler construction

The Danielewski–Fieseler examples of non-cancellation exploit the properties of the *Danielewski–Fieseler quotient.* Assume that the $\mathbb{G}_a$-action on X is free. Then the geometric orbit space $X/\mathbb{G}_a$ is a non-separated pre-variety (an algebraic space) obtained by gluing together several copies of $B := \operatorname{Spec} \mathcal{O}_X(X)^{\mathbb{G}_a}$ along a common Zariski open subset. The morphism μ can be factorized into $X \to X/\mathbb{G}_a \to B$. An ingenious observation by Danielewski is as follows. Consider two non-isomorphic smooth affine $\mathbb{G}_a$-surfaces X and Y with free $\mathbb{G}_a$-actions and with the same Danielewski–Fieseler quotient $F = X/\mathbb{G}_a = Y/G_a$. Then the affine threefold $W = X \times_F Y$ carries two induced free $\mathbb{G}_a$-actions. Moreover, W carries two different structures of principal $\mathbb{G}_a$-bundles (torsors) over X and over Y, respectively. Since X and Y are affine varieties, by Serre's Theorem ([**65**]) both these bundles are trivial, and so, $X \times \mathbb{A}^1 \cong W \cong Y \times \mathbb{A}^1$. This is exactly what happens for two different Danielewski surfaces $X = X_m$ and $Y = X_{m'}$, $m \neq m'$, and in other classical examples, see Section 3.5. The question arises as to how universal is the Danielewski-Fieseler construction. More precisely,

QUESTION. Let X and Y be non-isomorphic smooth affine surfaces with isomorphic cylinders $X \times \mathbb{A}^1 \cong Y \times \mathbb{A}^1$. Assume that both X and Y possess free $\mathbb{G}_a$-actions. Do there exist $\mathbb{A}^1$-fibrations on X and on Y over the same affine base and with the same Danielewski–Fieseler quotient?

Recall ([**19**, Def. 0.1]) that a *Danielewski-Fieseler surface* is a smooth affine surface X equipped with an $\mathbb{A}^1$-fibration $\mu: X \to \mathbb{A}^1$ which represents a (trivial) line bundle over $\mathbb{A}^1 \setminus \{0\}$ and such that the divisor $\mu^*(0)$ is reduced. Such a surface admits a free $\mathbb{G}_a$-action along the μ-fibers if and only if it is isomorphic to a surface in $\mathbb{A}^3$ with equation $xy - p(z) = 0$ where $p \in \Bbbk[z]$ has simple roots ([**19**, Cor.

4.13]). Theorem 5.2.2 below deals, more generally, with normal affine surfaces $\mathbb{A}^1$-fibered over affine curves and such that any fiber of the $\mathbb{A}^1$-fibration is reduced. Abusing the language we abbreviate these as *GDF-surfaces*, see Definition 2.1.1. The Danielewski trick does not work for them, in general, because such a surface does not need to admit a free $\mathbb{G}_a$-action. However, we show (see Theorems 5.2.2 and 7.2.1)

THEOREM 1.3.1. *A GDF-surface is a Zariski* 1*-factor if and only if it is the total space of a line bundle.*

The proof involves affine modifications, in particular, the *Asanuma modification.*

1.4. Affine modifications

Most of the known examples of non-cancellable affine surfaces exploit the Danielewski–Fieseler quotient, see, e.g., [**54, 69**]. By contrast, in this paper we use an alternative construction of non-cancellation due to T. Asanuma ([**6**]). Recall first the notion of an affine modification (see [**47**]).

DEFINITION 1.4.1 (*Affine modification*). Let $X = \operatorname{Spec}\mathfrak{A}$ be a normal affine variety where $\mathfrak{A} = \mathcal{O}_X(X)$ is the structure ring of X. Let further $I \subset \mathfrak{A}$ be an ideal, and let $f \in I$, $f \neq 0$. Consider the Rees algebra $\mathfrak{A}[tI] = \bigoplus_{n\geq 0} t^n I^n$ with $I^0 = \mathfrak{A}$ where t is an independent variable. Consider further the quotient $\mathfrak{A}' = \mathfrak{A}[tI]/(1-tf)$ by the principal ideal of $\mathfrak{A}[tI]$ generated by $1-tf$. The affine variety $X' = \operatorname{Spec}\mathfrak{A}'$ is called the *affine modification of* X *along divisor* $D = f^*(0)$ *with center* I. The inclusion $\mathfrak{A} \hookrightarrow \mathfrak{A}'$ induces a birational morphism $\varrho\colon X' \to X$ which contracts the exceptional divisor $E = (f\circ\varrho)^{-1}(0)$ on X' to the center $\mathbb{V}(I) \subset X$. In fact, any birational morphism of affine varieties $X' \to X$ is an affine modification ([**47**, Thm. 1.1]).

REMARKS 1.4.2. 1. If $I = (\mathfrak{a}_1, \dots, \mathfrak{a}_l)$ where $\mathfrak{a}_i \in \mathfrak{A}$, $i = 1, \dots, l$ then $\mathfrak{A}' = \mathfrak{A}[I/f] = \mathfrak{A}[\mathfrak{a}_1/f, \dots, \mathfrak{a}_l/f]$.

2. Assume that $f \in I_1 \subset I$ where I_1 is an ideal of $\mathfrak{A}$. Letting $\mathfrak{A}_1 = \mathfrak{A}[I_1/f]$ one obtains the equality $\mathfrak{A}' = \mathfrak{A}_1[I_2/f]$ where I_2 is the ideal generated by I in $\mathfrak{A}_1$. The inclusion $\mathfrak{A} \hookrightarrow \mathfrak{A}_1 \hookrightarrow \mathfrak{A}'$ leads to a factorization of the morphism $X' \to X$ into a composition of affine modifications, that is, birational morphisms of affine varieties $X' \to X_1 \to X$ where $X_1 = \operatorname{spec}\mathfrak{A}_1$ (cf. also [**47**, Prop. 1.2] for a different kind of factorization).

3. Geometrically speaking, the variety $X' = \operatorname{Spec}\mathfrak{A}'$ is obtained via blowing up $X = \operatorname{Spec}\mathfrak{A}$ at the ideal $I \subset \mathfrak{A}$ and deleting a certain transform of the divisor $D = f^*(0)$ on X', see [**47**] for details. However, in general $\mathbb{V}(I)$ might have components of codimension 1 which are then also components of the divisor $f^*(0)$. These components survive the modification. Thus, it is worth to distinguish between a *geometric* affine modification and an *algebraic* one.

Indeed, given a birational morphism of affine varieties $\sigma\colon X' \to X$ with exceptional divisor $E \subset X'$ and center $C = \sigma_*(E)$ of codimension at least 2, the divisor D of the associated modification can be defined as the closure of $X \smallsetminus \sigma(X')$ in X. However, this D is not necessarily a principal divisor. So, in order to represent $\sigma\colon X' \to X$ via an affine modification one needs to find a principal divisor on X with support containing D. Thus, although the data (D, C) is uniquely defined for

σ, there are many different affine modifications which induce the same birational morphism $\sigma: X' \to X$ (cf. [**21**] and also Remark 2.4.2 for the case of $\mathbb{A}^1$-fibered affine surfaces).

The following lemma will be used on several occasions. It generalizes [**47**, Cor. 2.2] with a similar proof.

LEMMA 1.4.3. *Let $X' \to X$ and $Y' \to Y$ be affine modifications along principal divisors $D_X = \operatorname{div} f_X$ and $D_Y = \operatorname{div} f_Y$ with centers I_X and I_Y, respectively, where $f_X \in I_X \setminus \{0\}$ and $f_Y \in I_Y \setminus \{0\}$. If an isomorphism $\varphi: X \xrightarrow{\cong} Y$ sends f_Y to f_X (hence, D_X to D_Y) and I_Y onto I_X then φ admits a lift to an isomorphism $\varphi': X' \xrightarrow{\cong} Y'$.*

We need also the following version of this lemma.

LEMMA 1.4.4. *Let X and Y be affine varieties, and let $\sigma: X \to Y$ be an affine modification along a principal divisor $\mathcal{D} = f^*(0)$ in Y with center an ideal $I \subset \mathcal{O}_Y(Y)$ where $f \in I \setminus \{0\}$. Let $\alpha \in \operatorname{Aut} Y$ be such that $\alpha(f) = f$ and both α, α^{-1} induce the identity on the sth infinitesimal neighborhood of $\mathcal{D}$ for some $s \geq 1$, that is,*

$$\alpha \equiv \mathrm{id} \mod f^s \quad \textit{and} \quad \alpha^{-1} \equiv \mathrm{id} \mod f^s .$$

Then α can be lifted to an automorphism $\tilde{\alpha} \in \operatorname{Aut} X$ such that

$$\tilde{\alpha} \equiv \mathrm{id} \mod f^{s-1} \quad \textit{and} \quad \tilde{\alpha}^{-1} \equiv \mathrm{id} \mod f^{s-1} . \tag{1.1}$$

PROOF. Let $\mathfrak{A} = \mathcal{O}_Y(Y)$ and $\mathfrak{A}' = \mathcal{O}_X(X) = \mathfrak{A}[\mathfrak{a}_1/f, \ldots, \mathfrak{a}_l/f]$ where $\mathfrak{a}_1, \ldots, \mathfrak{a}_l$ are generators of I. One has $\alpha^*(\mathfrak{a}_i) - \mathfrak{a}_i \in (f^s)$, that is, $\alpha^*(\mathfrak{a}_i) = \mathfrak{a}_i + f^s \mathfrak{b}_i$ for some $\mathfrak{b}_i \in \mathfrak{A}$, $i = 1, \ldots, l$. Extending α^* to an automorphism of the fraction field $\operatorname{Frac} \mathfrak{A}$ one has $\alpha^*(\mathfrak{a}_i/f) = \mathfrak{a}_i/f + f^{s-1}\mathfrak{b}_i$, $i = 1, \ldots, l$. Thus, $\alpha^*(\mathfrak{A}') \subset \mathfrak{A}'$ and, similarly, $(\alpha^{-1})^*(\mathfrak{A}') \subset \mathfrak{A}'$. So, $\tilde{\alpha}^* := \alpha^*|_{\mathfrak{A}'} \in \operatorname{Aut} \mathfrak{A}'$ yields an automorphism $\tilde{\alpha}$ of X verifying (1.1). □

It is easily seen that the affine modification of the linear space $\mathbb{A}^n$ with center in a linear subspace of codimension ≥ 2 and with divisor a hyperplane is isomorphic to $\mathbb{A}^n$. Similarly, certain affine *Asanuma modifications* of a cylinder give again a cylinder. This simple and elegant fact is due to Asanuma ([**6**]); we follow here [**45**, Lem. 7.9].

LEMMA 1.4.5. *Let X be an affine variety, $D = \operatorname{div} f$ a principal effective divisor on X where $f \in \mathcal{O}_X(X) \setminus \{0\}$, and $I \subset \mathcal{O}_X(X)$ an ideal with support contained in D. Let $X' \to X$ be the affine modification of X along D with center I. Consider the cylinder $\mathcal{X} = X \times \mathbb{A}^1 = \operatorname{Spec} \mathcal{O}_X(X)[v]$, the divisor $\mathcal{D} = D \times \mathbb{A}^1$ on $\mathcal{X}$, the ideal $\tilde{I} \subset \mathcal{O}_{\mathcal{X}}(\mathcal{X})$ generated by I, and the ideal $J = (\tilde{I}, v) \subset A[v]$ supported on $D \times \{0\} \subset \mathcal{D}$. Then the affine modifications of $\mathcal{X}$ along $\mathcal{D}$ with center $\tilde{I}$ and with center J are both isomorphic to the cylinder $\mathcal{X}' = X' \times \mathbb{A}^1$.*

PROOF. The affine modification of $\mathcal{X}$ along $\mathcal{D}$ with center $\tilde{I}$ yields the cylinder $\mathcal{X}'$. Let $\mathfrak{a}_1, \ldots, \mathfrak{a}_l \in I$ be generators of I, see 1.4.1 and 1.4.2. Then

$$\mathcal{O}_{\mathcal{X}'}(\mathcal{X}') = \mathcal{O}_X(X)[\mathfrak{a}_1/f, \ldots, \mathfrak{a}_l/f, v] \cong \mathcal{O}_X(X)[\mathfrak{a}_1/f, \ldots, \mathfrak{a}_l/f, v'/f] = \mathcal{O}_{\mathcal{X}''}(\mathcal{X}''),$$

where $v' = vf$ is a new variable, and $\mathcal{X}'' \to \mathcal{X}$ is the affine modification of $\mathcal{X}$ along $\mathcal{D}$ with center J. This gives the desired isomorphism. □

CHAPTER 2

$\mathbb{A}^1$-fibered surfaces via affine modifications

2.1. Covering trick and GDF surfaces

Throughout the paper we deal with the following class of $\mathbb{A}^1$-fibered surfaces.

DEFINITION 2.1.1 (*a GDF surface*). Let X be a normal affine surface over $\Bbbk$. A morphism $\pi: X \to B$ onto a smooth affine curve B is called an *$\mathbb{A}^1$-fibration* if the fiber $\pi^{-1}(b)$ over a general point $b \in B$ is isomorphic to the affine line $\mathbb{A}^1$ over $\Bbbk$. An $\mathbb{A}^1$-fibered surface $\pi: X \to B$ is called a *generalized Danielewski-Fieseler surface*, or a *GDF surface* for short, if all the fibers $\pi^*(b)$, $b \in B$, are reduced. In the case where $B = \mathbb{A}^1$ and $\pi^{-1}(0)$ is the only reducible fiber of π such surfaces were studied in [**19**] under the name *Danielewski-Fieseler surfaces*.

Any GDF surface is smooth, see, e.g., [**19**] or Lemma 2.3.8(b) below.

We say that a GDF surface $\pi: X \to B$ is *marked* if a *marking* $z \in \mathcal{O}_B(B) \smallsetminus \{0\}$ is given such that $z \circ \pi \in \mathcal{O}_X(X)$ vanishes to order one along any degenerate fiber of π. Abusing notation we often view z as a function on X identifying z and $z \circ \pi$. The components of the divisor $z^*(0)$ will be called *special fiber components*.

A GDF surface $\pi: X \to B$ equipped with actions of a finite group G on X and on B making the morphism π G-equivariant is called a *GDF G-surface*. Assuming that $G = \mu_d$ is the group of dth roots of unity, we choose a μ_d-quasi-invariant marking $z \in \mathcal{O}_B(B)$ of weight 1, that is, which satisfies $\zeta.z = \zeta \cdot z$ $\forall \zeta \in \mu_d$. Then we say that $\pi: X \to B$ is a *marked GDF μ_d-surface*.

Lemma 2.1.3 below is well known; for the sake of completeness we indicate a proof. This lemma says that, starting with a normal affine $\mathbb{A}^1$-fibered surface and applying a suitable cyclic Galois base change, it is possible to obtain a marked GDF μ_d-surface. The proof uses the following branched covering construction.

DEFINITION 2.1.2 (*Branched covering construction*). Consider a normal affine $\mathbb{A}^1$-fibered surface $\pi': Y \to C$ over a smooth affine curve C. Fix a finite set of points $p_1, \ldots, p_t \in C$ such that for any $p \in C \smallsetminus \{p_1, \ldots, p_t\}$ the fiber $\pi'^*(p)$ is reduced and irreducible. Let d be the least common multiple of the multiplicities of the components of the divisor $\sum_{i=1}^t \pi'^*(p_i)$ on Y. Choose a regular function $h \in \mathcal{O}_C(C)$ with only simple zeros which vanishes in the points $p_1, \ldots, p_t$ and eventually somewhere else. Letting $\mathbb{A}^1 = \operatorname{spec} \Bbbk[z]$ consider the smooth curve $B \subset C \times \mathbb{A}^1$ given by equation $z^d - h(p) = 0$ where $(p, z) \in C \times \mathbb{A}^1$ along with the morphism $\mathrm{pr}_1: B \to C$. By abuse of notation we denote the function $z|_B \in \mathcal{O}_B(B)$ still by z. Let X be the normalization of the cross-product $Y \times_C B$, and let $\pi: X \to B$ and $\varphi: X \to Y$ be the induced morphisms.

LEMMA 2.1.3. *In the notation of* 2.1.2 *the following holds.*

- *The cyclic group μ_d of order d acts naturally on B so that $C = B/\mu_d$;*

- *the morphism* $\mathrm{pr}_1: B \to C$ *is ramified to order* d *over the zeros of* h. *The function* $z \in \mathcal{O}_B(B)$ *is a* μ_d*-quasi-invariant of weight* 1, *and* $\operatorname{div} z = d^{-1}\mathrm{pr}_1^*(\operatorname{div} h)$ *is a reduced effective* μ_d*-invariant divisor on* B;
- *the morphism* $\varphi: X \to Y$ *of* $\mathbb{A}^1$*-fibrations is a cyclic covering with the Galois group* μ_d, *the reduced branching divisor* $h^*(0)$ *on* Y, *and the ramification divisor* $z^*(0)$ *on* X;
- *the* μ_d*-equivariant morphism* $\pi: X \to B$ *and the marking* $z \in \mathcal{O}_B(B)$ *define a structure of a marked GDF* μ_d*-surface on* X.

PROOF. The map $\nu_d: \mathbb{A}^1 \to \mathbb{A}^1$, $z \mapsto z^d$, is the quotient morphism of the natural μ_d-action on $\mathbb{A}^1$. The first three statements follow from the fact that the curve B along with the morphism $z: B \to \mathbb{A}^1$ is obtained using the morphism $h: C \to \mathbb{A}^1$ via the base change $\nu_d: \mathbb{A}^1 \to \mathbb{A}^1$ that fits in the commutative diagram

$$(2.1) \qquad \begin{array}{ccc} X & \xrightarrow{/\mu_d} & Y \\ {\scriptstyle \pi}\downarrow & & \downarrow{\scriptstyle \pi'} \\ B & \xrightarrow{/\mu_d} & C \\ {\scriptstyle z}\downarrow & & \downarrow{\scriptstyle h} \\ \mathbb{A}^1 & \xrightarrow[\nu_d]{} & \mathbb{A}^1 \end{array}$$

The remaining assertions can be reduced to a simple computation in local charts. Indeed, let (t, u) be coordinates in a local analytic chart U in Y centered at a smooth point $y \in Y$ which is a general point of a fiber component F over p_i of multiplicity n in the divisor $(\pi')^*(p_i)$. We may choose t so that $h \circ \pi'|_U = t^n$ and $F \cap U = t^*(0)$. Then $Y \times_C B$ is given locally in $\mathbb{A}^3$ with coordinates (z, t, u) by equation $z^d - t^n = 0$ where $n|d$ by our choice of d. This is a union of n smooth surface germs $\{z^{d/n} - \zeta t = 0\}$ where $\zeta^n = 1$, meeting transversely along the line $z = t = 0$ that projects in Y onto $F \cap U$. Passing to a normalization one gets n smooth disjoint surface germs, say, $V_1, \dots, V_n$ in X over U. The function $z \in \mathcal{O}_X(X)$ gives in each chart V_j a local coordinate such that $\varphi^*(F) = z^*(0)$ has multiplicity one in V_j. We leave the further details to the reader. □

2.1.4 (*Cancellation Problem for surfaces: a reduction*). The following reasoning is borrowed in [**54, 55, 68**]. It occurs that in order to construct (families of) $\mathbb{A}^1$-fibered surfaces with isomorphic cylinders it suffices to construct (families of) $\mathbb{A}^1$-fibered GDF G-surfaces with G-equivariantly isomorphic cylinders.

Suppose that a Galois base change $B \to C$ with a Galois group G applied to two distinct $\mathbb{A}^1$-fibered surfaces $\pi'_j: Y_j \to C$, $j = 0, 1$, yields two $\mathbb{A}^1$-fibered GDF G-surfaces $\pi_j: X_j \to B$, $j = 0, 1$, with G-isomorphic over B cylinders $X_0 \times \mathbb{A}^1 \cong_{G,B} X_1 \times \mathbb{A}^1$ where in the both cases G acts identically on the second factor $\mathbb{A}^1$. Clearly, one has $(X_j \times \mathbb{A}^1)/G \cong Y_j \times \mathbb{A}^1$, $j = 0, 1$. Passing to the quotients yields an isomorphism over C of cylinders $Y_0 \times \mathbb{A}^1 \cong_C Y_1 \times \mathbb{A}^1$ that fits in the commutative diagram

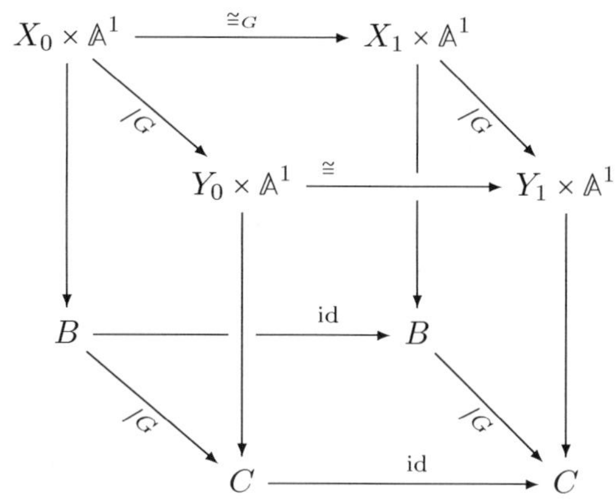

In the sequel we will concentrate on the following problem. Consider the cylinders $X \times \mathbb{A}^1$ and $X' \times \mathbb{A}^1$ over two $\mathbb{A}^1$-fibered GDF surfaces $\pi: X \to B$ and $\pi': X' \to B$ with the same smooth affine base B. Suppose that π and π' are equivariant with respect to actions of a finite group G on X, X', and B. We extend these actions to G-actions on the cylinders $X \times \mathbb{A}^1$ and $X' \times \mathbb{A}^1$ identically on the second factor.

PROBLEM 2.1.5. *Find a criterion for two GDF G-surfaces $\pi_X: X \to B$ and $\pi_{X'}: X' \to B$ over the same base B as to when the cylinders $X \times \mathbb{A}^1$ and $X' \times \mathbb{A}^1$ are G-equivariantly isomorphic.*

In Theorems 5.2.2, 5.3.1 and Propositions 7.1.14 and 7.1.17 we provide some sufficient conditions in the case where $G = \mu_d$. Actually, these conditions guarantee the existence of a G-equivariant isomorphism $X \times \mathbb{A}^1 \overset{\cong_{B,G}}{\longrightarrow} X' \times \mathbb{A}^1$ which respects the natural projections $X \times \mathbb{A}^1 \to B$ and $X' \times \mathbb{A}^1 \to B$ and induces the identity on B.

2.2. Pseudominimal completion and extended divisor

DEFINITION 2.2.1 (*Pseudominimal resolved completion*). Any $\mathbb{A}^1$-fibration $\pi: X \to B$ on a normal affine surface X over a smooth affine curve B extends to a $\mathbb{P}^1$-fibration $\tilde{\pi} : \tilde{X} \to \bar{B}$ on a complete surface $\tilde{X}$ over a smooth completion $\bar{B}$ of B such that $D = \tilde{X} \setminus X$ is a simple normal crossing divisor carrying no singular point of $\tilde{X}$. Let $\varrho : \bar{X} \to \tilde{X}$ be the minimal resolution of singularities (all of these singularities are located in X). Abusing notation, we consider D as a divisor in $\bar{X}$. We call $(\bar{X}, D)$ a *resolved completion* of X.

Consider the induced $\mathbb{P}^1$-fibration $\bar{\pi} := \tilde{\pi} \circ \varrho : \bar{X} \to \bar{B}$. There is a unique (*horizontal*) component S of D which is a section of $\bar{\pi}$, while all the other (*vertical*) components of D are fiber components. Let $\bar{B} \setminus B = \{c_1, \dots, c_s\}$. Contracting subsequently the (-1)-components of D different from S we may assume in addition that D does not contain any (-1)-component of a fiber. Such a resolved completion $(\bar{X}, D)$ is called *pseudominimal.*

DEFINITION 2.2.2 (*Extended divisor*). Let $(\bar{X}, D)$ be a resolved completion of X along with the associate $\mathbb{P}^1$-fibration $\bar{\pi}: \bar{X} \to \bar{B}$, and let $b_1, \dots, b_n$ be the points of B such that the fibers $\bar{\pi}^*(b_i)$ over b_i in $\bar{X}$ are degenerate, i.e., are either non-reduced or reducible. The reduced divisor

$$D_{\text{ext}} = D \cup \Lambda \quad \text{where} \quad \Lambda = \bigcup_{j=1}^{n} \bar{\pi}^{-1}(b_j) \tag{2.2}$$

is called the *extended divisor* of $(\bar{X}, D)$, and the weighted dual graph Γ_{ext} of D_{ext} the *extended graph* of $(\bar{X}, D)$. Recall that the vertices of Γ_{ext} correspond to the components of D_{ext}, and the weight attributed to a vertex is the selfintersection index of the corresponding component. We say that Γ_{ext} is *pseudominimal* if the completion $(\bar{X}, D)$ is. The graph Γ_{ext} is a rooted tree with the horizontal section $S \subset D$ as a root. The dual graph $\Gamma(D)$ of the boundary divisor D is a rooted subtree of Γ_{ext}.

For a subgraph Γ' of a graph Γ we let $\Gamma \ominus \Gamma'$ denote the graph obtained from Γ by deleting the vertices of Γ' along with all their incident edges of Γ. The connected components of $\Gamma_{\text{ext}} \ominus \Gamma(D)$ are called the *feathers* of D_{ext}. Under the pseudominimality assumption all the (-1)-components of Λ are among the feather components.

DEFINITION 2.2.3 (*Standard completion*). Consider a pseudominimal resolved completion $\bar{\pi}: \bar{X} \to \bar{B}$. The fibers $\bar{\pi}^{-1}(c_i)$ where $c_i \in \bar{B} \setminus B$, $i = 1, \ldots, s$ are reduced and irreducible 0-curves. Performing, if necessary, elementary transformations in one of them we may assume that also the section S is a 0-curve. Such a completion will be called *standard*, cf., e.g., [**31**, 5.11]. By [**31**, Lem. 5.12], if two $\mathbb{A}^1$-fibrations $\pi: X \to B$ and $\pi': X' \to B$ are isomorphic over B then the corresponding standard extended divisors D_{ext} and D'_{ext} and the corresponding (unweighted) extended graphs Γ_{ext} and Γ'_{ext} are.

REMARK 2.2.4 (*Fiber structure*). Recall (see [**56**, Ch. 3, Lem. 1.4.1 and 1.4.4]) that any degenerate fiber of $\pi: X \to B$ is a disjoint union of components isomorphic to $\mathbb{A}^1$, any singular point of X is a cyclic quotient singularity, and two such singular points cannot belong to the same component. The minimal resolution of a singular point has as exceptional divisor in $\bar{X}$ a chain of rational curves with no (-1)-component and with a negative definite intersection form. This chain meets just one other fiber component at a terminal component of the chain.

DEFINITION 2.2.5 (*Bridges*). Any feather $\mathfrak{F}$ of D_{ext} (see 2.2.2) is a chain of smooth rational curves on $\tilde{X}$ with dual graph

$$\Gamma(\mathfrak{F}): \quad \overset{F_0}{\circ}\!\!-\!\!\!-\!\!\!-\!\!\overset{F_1}{\circ}\!\!-\!\!\!-\ \cdots\ -\!\!\!-\!\!\overset{F_k}{\circ}\ .$$

The subchain $\mathfrak{R} = \mathfrak{F} \ominus F_0 = F_1 + \ldots + F_k$ (if non-empty) contracts to a cyclic quotient singularity of X. The component F_0 called the *bridge* of $\mathfrak{F}$ is attached to a unique component C of D. The bridge F_0 is the closure in $\bar{X}$ of a fiber component $F_0 \setminus C \cong \mathbb{A}^1$ of π. Vice versa, for each fiber component F of π the closure $\bar{F} \subset \bar{X}$ of the proper transform of F is a bridge of a unique feather. In the case of a smooth surface X one has $k = 0$, i.e., any feather $\mathfrak{F}$ consists of a bridge: $\mathfrak{F} = F_0$.

2.3. Blowup construction

DEFINITION 2.3.1 (*Blowup construction*). Let $\pi: X \to B$ be an $\mathbb{A}^1$-fibration on a normal affine surface X over a smooth affine curve B, and let $(\bar{X}, D)$ be a resolved completion of X along with the associate $\mathbb{P}^1$-fibration $\bar{\pi}: \bar{X} \to \bar{B}$ and with a section 'at infinity' S. In any degenerate fiber $\bar{\pi}^*(b_i)$ on $\bar{X}$, $i = 1, \ldots, n$, there is a unique component, say, C_i meeting S. The next fact is well known. For the reader's convenience we provide a brief argument.

LEMMA 2.3.2. *Let C_0 be the component of a reducible fiber $\bar\pi^{-1}(b)$, $b \in B$, such that $C_0 \cdot S = 1$. Then the rest of the fiber $\bar\pi^{-1}(b) \ominus C_0$ can be blown down to a smooth point.*

PROOF. Since $S \cdot \bar\pi^*(b) = S \cdot C_0 = 1$, C_0 has multiplicity 1 in the fiber. We proceed by induction on the number N of components in the fiber $\bar\pi^{-1}(b)$. The statement is clearly true for $N = 1$. Suppose now that $N > 1$. Then there exists a (-1)-component E in the fiber. If $E \neq C_0$ then contracting E one can use the induction hypothesis. Assume now that C_0 is the only (-1)-component of $\bar\pi^{-1}(b)$. Since C_0 has multiplicity 1 it can be contracted to a smooth point of the resulting fiber sitting on a multiplicity 1 component, say, C_1. By the induction hypothesis after blowing down C_0 the rest of the resulting fiber but C_1 can be blown down. Thus there is a (-1)-component of the fiber $\bar\pi^{-1}(b)$ disjoint from C_0. However, the latter contradicts our assumption that C_0 is a unique (-1)-component of the fiber $\bar\pi^{-1}(b)$. □

Performing such a contraction for every $i = 1, \ldots, n$ one arrives at a geometrically minimal ruling (that is, a locally trivial $\mathbb{P}^1$-fibration) $\bar\pi_0 : \bar X_0 \to \bar B$. The image $S_0 \subset \bar X_0$ of S is a section of $\bar\pi_0$. Thus $\bar X$ can be obtained starting with a geometrically ruled surface $\bar X_0$ via a sequence of blowups of points

$$\bar X = \bar X_m \xrightarrow{\bar\varrho_m} \bar X_{m-1} \longrightarrow \ldots \longrightarrow \bar X_1 \xrightarrow{\bar\varrho_1} \bar X_0 \tag{2.3}$$

with centers contained in $\bar\pi_0^{-1}(b_i) \smallsetminus S_0 \subset \bar X_0$, $i = 1, \ldots, n$, and at infinitely near points. For $j = 0, \ldots, m$ we let $\bar\pi_j : \bar X_j \to \bar B$ be the induced $\mathbb{P}^1$-fibrations.

DEFINITION 2.3.3 (*Well ordered blowup construction*). In the rooted tree Γ_{ext} with a root S, the (-1)-vertices on the maximal distance from S are disjoint from S and mutually disjoint due to Lemma 2.3.2. Hence the corresponding fiber components can be simultaneously contracted. Repeating this procedure one arrives finally at a smooth geometrically ruled surface $\bar\pi_0 : \bar X_0 \to \bar B$ along with a specific sequence (2.3) of blowups where every $\bar\varrho_i$, $i = 1, \ldots, n$, is a blowup with center in a reduced zero dimensional subscheme of $\bar X_{i-1} \smallsetminus (\bar\pi_{i-1}^{-1}(\bar B \smallsetminus B) \cup S_{j-1})$ where S_j is the proper transform on $\bar X_j$ of $S_0 \subset \bar X_0$. We call such a sequence (2.3) a *well ordered blowup construction.*

The following lemma is a generalization of Theorem 2.1 in [**26**].

LEMMA 2.3.4. *Let $\pi : X \to B$ be an $\mathbb{A}^1$-fibered GDF G-surface where G is a finite group. Then there is a G-equivariant resolved completion $(\bar X, D)$ of X obtained via a G-equivariant well ordered blowup construction* (2.3).

PROOF. By Sumihiro Theorem ([**67**, Thm. 3]) there exists a G-equivariant projective completion $(\tilde X, \tilde D)$ of X. The minimal resolution of singularities of the pair $(\tilde X, \tilde D)$ is G-equivariant (being unique). In this way we arrive at a G-equivariant smooth projective completion $(\bar X, D)$ of X with a G-invariant simple normal crossing divisor D. The closures in $\bar X$ of the fibers of $\pi : X \to B$ form a (nonlinear) G-invariant pencil. Its base points also admit a G-equivariant resolution. Hence we may assume that $\bar X$ comes equipped with a G-equivariant $\mathbb{P}^1$-fibration $\bar\pi : \bar X \to \bar B$ along with a G-invariant section S of $\bar\pi$.

In particular, the root S of the extended graph Γ_{ext} of $(\bar{X}, D)$ is fixed by the induced G-action on Γ_{ext}. This action stabilizes as well the set of (-1)-vertices on the maximal distance from S. Therefore, the simultaneous contraction of the corresponding fiber components is G-equivariant. By recursion one arrives at a G-equivariant well ordered blowup construction. □

REMARKS 2.3.5. 1. Under a well ordered blowup construction (2.3) no blowup is done near the section at infinity S_0 of $\bar{\pi}_0$ neither with center over the points $c_i \in \bar{B} \setminus B$, $i = 1, \ldots, k$. So, the fibers in $\bar{X}_i$ over these points remain reduced and irreducible.

2. Let a component F of D_{ext} different from S be created by one of the blowups $\bar{\varrho}_\nu : \tilde{X}_\nu \to \tilde{X}_{\nu-1}$ in (2.3). We claim that then the center P_ν of the blowup $\bar{\varrho}_\nu$ belongs to the image of D in $\bar{X}_{\nu-1}$. Indeed, otherwise the last (-1)-curve, say, E over P_ν would neither be a bridge of a feather, nor a component of D. Hence E should be a component of a feather, say, $\mathfrak{F}$, different from the bridge component F_0. However, the latter contradicts the minimality of $\mathfrak{F} \ominus F_0$, that is, the minimality of the resolution of singularities of X.

Recall the following notions.

2.3.6. Let D be a simple normal crossing divisor on a smooth surface Y. A blowup of Y at a point $p \in D$ is called *outer* if p is a smooth point of D and *inner* if p is a node.

We use the following notation.

NOTATION 2.3.7. Given a blowup construction (2.3) for any $j = 0, \ldots, m$ we let

$$D_{j,\text{ext}} = S_j \cup \Delta_j \cup \Lambda_j \subset \bar{X}_j \quad \text{where} \quad \Delta_j = \bigcup_{i=1}^{k} \bar{\pi}_j^{-1}(c_i) \text{ and } \Lambda_j = \bigcup_{i=1}^{n} \bar{\pi}_j^{-1}(b_i). \tag{2.4}$$

The following lemma should be well known. For (b) see, e.g., [**19**, (2.2)] and the proof of Proposition 6.3.23 in [**32**].

LEMMA 2.3.8. *Let $\pi: X \to B$ be a normal affine $\mathbb{A}^1$-fibered surface over a smooth affine curve B. Consider a resolved completion $(\bar{X} = \bar{X}_m, D)$ of X obtained via a well ordered blowup construction (2.3) starting with a ruled surface $\bar{\pi}_0: \bar{X}_0 \to \bar{B}$. Then the following hold.*

(a) *$\pi: X \to B$ is a GDF surface if and only if all the blowups $\bar{\varrho}_\nu$ in (2.3), $\nu = 1, \ldots, m$, are outer (with respect to the divisor $D_{0,\text{ext}}$ on $\bar{X}_0$ and its subsequent total transforms $D_{\nu,\text{ext}}$ on $\bar{X}_\nu$).*

(b) *If $\pi: X \to B$ is a GDF surface then X is smooth and every feather $\mathfrak{F}$ of $D_{\text{ext}} = D_{m,\text{ext}}$ consists of a single (-1)-component F_0 which is a bridge.*

(c) *Let $\pi: X \to B$ be a GDF surface with a pseudominimal resolved completion $(\bar{X}, D)$, see Definition 2.2.1. For a fiber component F of π the following are equivalent:*
- *$\bar{F}$ is a leave, that is, an extremal vertex of the rooted tree Γ_{ext};*
- *$\bar{F}$ is a feather;*
- *$\bar{F}$ is a (-1)-vertex of Γ_{ext}.*

PROOF. Suppose that for some $\nu \in \{1, \ldots, m\}$ the blowup $\bar{\varrho}_\nu$ is inner. Assume also that the center $P_\nu \in \bar{X}_{\nu-1}$ of $\bar{\varrho}_\nu$ lies on the fiber over $b_i \in B$ and on the image $D_{\nu-1,\text{ext}}$ of D_{ext}. Then all components of the fiber $\bar{\pi}^*(b_i)$ which are born over P_ν

including the last (-1)-component, say, $\bar{F}$, have multiplicities > 1. Notice that $\bar{F} = \bar{F}_0$ is a bridge component of a feather, say, $\mathfrak{F}$. Hence $\bar{F}$ is the closure in $\bar{X}$ of a component F of the fiber $\pi^*(b_i) \subset X$. Thus, the fiber $\pi^*(b_i)$ is not reduced. This contradiction shows that for a GDF surface $\pi: X \to B$ all the $\bar{\varrho}_\nu$, $\nu = 1, \ldots, m$, are outer.

To show the converse suppose that all the $\bar{\varrho}_\nu$ in (2.3), $\nu = 1, \ldots, m$ are outer. Then all the resulting degenerate fibers are reduced. Hence $\pi: X \to B$ is a GDF surface. This proves (a).

Assume further that a feather $\mathfrak{F}$ of D_{ext} has more than one component. The component of $\mathfrak{F}$ which appears the last in the blowup construction (2.3) is the bridge component $\bar{F}_0$ of $\mathfrak{F}$. Hence $\bar{F}_0$ results from a blowup $\bar{\varrho}_\nu$ with center P_ν which lies on the component $\bar{F}_1$ of $\mathfrak{F}$ and on the image in $\bar{X}_{\nu-1}$ of a component C of D, see Remark 2.3.5. Thus, P_ν is a nodal point of the divisor $D_{\nu-1,\text{ext}}$ on $\bar{X}_{\nu-1}$. It follows that $\bar{\varrho}_\nu$ is inner. So, the bridge $\bar{F}_0$ of $\mathfrak{F}$ has multiplicity > 1 in its fiber.

This proves that for a GDF surface $\pi: X \to B$ every feather $\mathfrak{F}$ of D_{ext} consists of a single bridge component $\bar{F}_0$. Consequently, the surface X is smooth. Furthermore, assuming that $\bar{F}_0^2 < -1$ an outer blowup was done in (2.3) with center on $\bar{F}_0$ creating a new component, say, E of D. The graph distance $\text{dist}(E, S)$ in Γ_{ext} is bigger than $\text{dist}(\bar{F}_0, E)$. Hence $\bar{F}_0$ disconnects S and E in D. The latter contradicts the facts that the affine surface X is connected at infinity, i.e., its boundary divisor D is connected. Therefore, $\bar{F}_0^2 = -1$. This shows (b).

The same argument shows that $\bar{F}_0$ is an extremal vertex (a *tip*) of Γ_{ext} different from S. Conversely, if $\bar{F}$ is a tip of Γ_{ext} different from S then $\bar{F}^2 = -1$. Indeed, since all the blowups in (2.3) are outer then no further blowup was done near $\bar{F}$ after creating $\bar{F}$. Due to the pseudominimality assumption, $\bar{F}$ is a feather of D_{ext}. Now (c) follows. □

DEFINITION 2.3.9 (*Fiber trees, levels, and types*). Given an SNC completion $\bar{\pi}: \bar{X} \to \bar{B}$ of a GDF surface $\pi: X \to B$ and a point $b \in B$ the dual graph $\Gamma_b = \Gamma_b(\bar{\pi})$ of the fiber $\bar{\pi}^{-1}(b)$ will be called a *fiber tree*. It depends on the completion chosen. This is a rooted tree with a root $v_0 \in \Gamma_b$ being the neighbor of S in Γ_{ext}. We say that a vertex v of Γ_b has *level* l if the tree distance between v and v_0 equals l. Thus, the root v_0 is a unique vertex of Γ_b on level 0. By a *height* $\text{ht}(\Gamma_b)$ we mean the highest level of the vertices in Γ_b. Remind that the *leaves* of a rooted tree are its extremal vertices different from the root. By the *type* $\text{tp}(\Gamma_b)$ we mean the sequence of nonnegative integers $(n_1, n_2, \ldots, n_h)$ where $h = \text{ht}(\Gamma_b)$ and n_i is the number of leaves of Γ_b on level i.

REMARK 2.3.10. The fiber tree Γ_b is an unweighted tree. However, one can easily reconstruct the weights as defined in 2.2.2. Namely, for a vertex v of weight $w(v)$ and of degree $\deg(v)$ in Γ_b one has $w(v) = -\deg(v)$. In particular, the (-1)-vertices are the tips, and the (-2)-vertices are the linear ones.

DEFINITION 2.3.11 (*Graph divisor*). Let $\mathfrak{G}$ be the set of all finite weighted rooted trees contractible to the root which acquires then weight zero. By a *graph divisor* on a smooth affine curve B we mean a formal sum

$$\mathcal{D} = \sum_{i=1}^{n} \Gamma_i b_i, \quad \text{where} \quad \Gamma_i \in \mathfrak{G}.$$

If all the Γ_i are chains then we call $\mathcal{D}$ a *chain divisor*. The *height* of a graph divisor $\mathcal{D}$ is the maximal height of the trees Γ_i, $i = 1, \ldots, n$.

Let $\pi: X \to B$ be an $\mathbb{A}^1$-fibered surface with a marking $z \in \mathcal{O}_B(B)$ where $z^*(0) = b_1 + \ldots + b_n$ and with a resolved completion $\bar{\pi}: \bar{X} \to \bar{B}$. To the corresponding extended graph Γ_{ext} we associate a *graph divisor* $\mathcal{D}(\pi) = \sum_{i=1}^n \Gamma_{b_i} b_i$ where Γ_{b_i} is the fiber tree of the fiber $\pi^{-1}(b_i)$. If $\pi: X \to B$ is a μ_d-surface and the marking z is μ_d-quasi-invariant then there is an induced μ_d-action on the graph divisor $\mathcal{D}(\pi)$.

2.4. GDF surfaces via affine modifications

Let $X \to B$ be a GDF surface. In this section we describe a recursive procedure which allows to recover X starting with the product $B \times \mathbb{A}^1$ via a sequence of fibered modifications, see Corollary 2.4.6.

DEFINITION 2.4.1 (*Fibered modification*). [1] A *fibered modification* between two $\mathbb{A}^1$-fibered GDF surfaces $\pi: X \to B$ and $\pi': X' \to B$ is an affine modification $\varrho: X' \to X$ which consists in blowing up a reduced zero-dimensional subscheme of X and deleting the proper transform of the union of those fiber components of π which carry centers of blowups. Thus, ϱ is a birational morphism of B-schemes.

REMARK 2.4.2. Let F be a reduced curve on a smooth affine surface X, let Σ be a reduced principal divisor on F, and let $\sigma: X' \to X$ be the composition of blowing up X with center Σ and deleting the proper transform F' of F. We claim that X' is again affine, and so, by [**47**, Thm. 1.1], the birational morphism $X' \to X$ is an affine modification.

Indeed, there exists a completion $\bar{X}$ of X and an ample divisor A on $\bar{X}$ with support $\operatorname{supp} A = \bar{X} \setminus X$. Let $\bar{X}'$ be the surface obtained from $\bar{X}$ by blowing up with center Σ. Consider on $\bar{X}'$ the proper transforms A' and $\bar{F}'$ of A and $\bar{F}$, respectively, where $\bar{F}$ is the closure of F in $\bar{X}$. By Kleiman ampleness criterion the divisor $nA' + \bar{F}'$ on $\bar{X}'$ is ample provided that n is sufficiently large. Hence the surface $X' = \bar{X}' \setminus \operatorname{supp}(nA' + \bar{F}')$ is affine, as claimed.

In general, F is not a principal divisor on X and $\sum$ is not a principal divisor on F. Nonetheless, one can represent $\sigma: X' \to X$ via an affine modification along a principal divisor provided there exist a function $f \in \mathcal{O}_X(X)$ such that $f^*(0) = F + G$ is a reduced divisor and the support of G is disjoint with F. Indeed, let $I \subseteq \mathcal{O}_X(X)$ be the ideal of $\sum \cup G$. Then $\sigma: X' \to X$ is the affine modification along the divisor $f^*(0)$ with center I.

Let now $\pi: X \to B$ be a GDF surface, F be a fiber component of π, and let $f = \pi^* z$ where $z \in \mathcal{O}_B(B)$ has a simple zero at the point $\pi(F) \in B$. Then $\pi' = \pi \circ \sigma: X' \to B$ is again a GDF surface and $\sigma: X' \to X$ is a fibered modification. This justifies Definition 2.4.1.

For a GDF surface $\pi: X \to B$ one has the following decomposition.

PROPOSITION 2.4.3. (a) *Any GDF surface $\pi: X \to B$ can be obtained starting with a line bundle $\pi_0: X_0 \to B$ over B via a sequence of fibered modifications*

$$X = X_m \xrightarrow{\varrho_m} X_{m-1} \longrightarrow \ldots \longrightarrow X_1 \xrightarrow{\varrho_1} X_0 . \tag{2.5}$$

This sequence can be extended to the corresponding completions yielding a well ordered blowup sequence (2.3).

[1] Cf. [**19**, Def. 4.2].

(b) *Suppose, furthermore, that* $\pi: X \to B$ *is a GDF* G*-surface where* G *is a finite group. Then* (2.5) *can be chosen so that the intermediate surfaces* X_ν *come equipped with* G*-actions making the morphisms* $\varrho_{\nu+1}: X_{\nu+1} \to X_\nu$ *and* $\pi_\nu: X_\nu \to B$ G*-equivariant for all* $\nu = 0, \dots, m-1$.

PROOF. (a) To construct (2.5) we exploit a well ordered blowup construction (2.3) which starts with a $\mathbb{P}^1$-bundle $\bar{\pi}_0: \bar{X}_0 \to \bar{B}$ and finishes with a pseudominimal completion $\bar{\pi}_m: \bar{X}_m \to \bar{B}$ of $\pi: X \to B$.

For any $\nu = 0, \dots, m$ we let $D_{\nu,\mathrm{ext}}$ (Δ_ν, S_μ, respectively) be the image on $\bar{X}_\nu$ of the extended divisor $D_{\mathrm{ext}} = D_{m,\mathrm{ext}}$ (the divisor $\Delta = \Delta_m$, the section $S = S_m$, respectively) on $\bar{X}_m = \bar{X}$. Let $\Gamma_{\nu,\mathrm{ext}}$ be the weighted dual graph of $D_{\nu,\mathrm{ext}}$ and $\Lambda_{\nu,\max}$ be the union of the fiber components of $\bar{\pi}_\nu: \bar{X}_\nu \to \bar{B}$ which correspond to the extremal vertices of $\Gamma_{\nu,\mathrm{ext}}$ on maximal distance from S_ν. Let also D_ν be the union of the remaining components of $D_{\nu,\mathrm{ext}}$. Then $\Lambda_{\nu+1,\max}$ is the exceptional divisor of the blowup $\bar{\varrho}_{\nu+1}: \bar{X}_{\nu+1} \to \bar{X}_\nu$ with center on $\Lambda_{\nu,\max} \setminus D_\nu$.

Consider the open surface $X_\nu = \bar{X}_\nu \setminus D_\nu$. We claim that X_ν is affine and $\bar{\varrho}_{\nu+1}(X_{\nu+1}) \subset X_\nu$. Indeed, the latter follows since $\bar{\varrho}_{\nu+1}(\Lambda_{\nu+1,\max} \setminus D_{\nu+1}) \subset \Lambda_{\nu,\max} \setminus D_\nu \subset X_\nu$ due to the above observation. To prove the former we use the Kleiman ampleness criterion (cf. Remark 2.4.2). Notice that a fiber component F of D_{ext} has level l if $\mathrm{dist}(F, S) = l+1$ in Γ_{ext}. For any $\nu \ge l$ the proper transform of F in $\bar{X}_\nu$ has the same level l. Choose a sequence of positive integers

$$s_0 \gg a_0 \gg a_1 \gg \ldots \gg a_{m-1} \gg 0 .$$

Let A_ν be an effective divisor on $\bar{X}_\nu$ with support D_ν such that a_l is the multiplicity in A_ν of any fiber component F of D_ν of level l, $l = 0, \dots, \nu-1$, and s_0 is the multiplicity of S_ν in A_ν. Performing elementary transformations in a fiber over a point $c_i \in \bar{B} \setminus B$ we may assume that $S_\nu^2 > 0$. Suppose to the contrary that there is an irreducible curve C in $\bar{X}_\nu$ with $C \cdot A_\nu = 0$. If C is not a component of $D_{\nu,\mathrm{ext}}$ then $\bar{\pi}(C) \subset B$, hence the restriction $\bar{\pi}|_C$ is constant, which gives a contradiction. If $C = S_\nu$ then clearly $C \cdot A_\nu > 0$ which contradicts our choice of C and A_ν. The same contradiction occurs if C is a fiber component of $D_{\nu,\mathrm{ext}}$. Due to the Kleiman ampleness criterion the divisor A_ν with support D_ν is ample. Thus the surface $X_\nu = \bar{X}_\nu \setminus D_\nu$ is affine.

Letting now $\varrho_{\nu+1} = \bar{\varrho}_{\nu+1}|_{X_{\nu+1}}: X_{\nu+1} \to X_\nu$, $\nu = 0, \dots, m-1$ one obtains a desired sequence (2.5) of fibered modifications. This proves (a).

To show (b) it suffices to start with a G-equivariant version of sequence (2.3) constructed in the proof of Lemma 2.3.4(b). By our construction, $\bar{\varrho}_\nu$ is G-equivariant and $D_{\nu,\mathrm{ext}}$, D_ν, and X_ν are G-invariant. Hence $\varrho_{\nu+1} = \bar{\varrho}_{\nu+1}|_{X_{\nu+1}}$ is G-equivariant too for any $\nu = 0, \dots, m-1$. □

The following proposition is an affine analog of the Nagata-Maruyama Theorem on the projective ruled surfaces ([**60**]; see also [**49**]). It allows to replace the line bundle $X_0 \to B$ in (2.5) by the trivial bundle $B \times \mathbb{A}^1 \to B$. For the corresponding completions this amounts to a stretching which extends feathers with chains of type $[[-1, -2, \dots, -2]]$ in D near S, so loosing the pseudominimality property.

LEMMA 2.4.4. *Let* X *be the total space of a line bundle* $\pi: X \to B$ *over a smooth affine curve* B. *Then the following hold.*

(a) *X can be obtained starting with the product $B \times \mathbb{A}^1$ over B via a sequence of fibered modifications over B,*

$$X = Z_M \xrightarrow{\varrho_M} Z_{M-1} \longrightarrow \ldots \longrightarrow Z_1 \xrightarrow{\varrho_1} Z_0 = B \times \mathbb{A}^1 \tag{2.6}$$

where $\pi_i: Z_i \to B$ is the projection of a line bundles and the center of ϱ_i belongs to the exceptional divisor of ϱ_{i-1} for each $i = 0, \ldots, M$.

(b) *If in addition $\pi: X \to B$ is a marked GDF μ_d-surface then for $i = 0, \ldots, M$ the morphisms $\varrho_i: Z_i \to Z_{i-1}$ in (2.6) and $\pi_i: Z_i \to B$ can be chosen to be μ_d-equivariant with respect to suitable μ_d-actions on the surfaces Z_i and the given μ_d-action on B.*

PROOF. (a) Under our assumptions X is affine and admits an effective $\mathbb{G}_m$-action along the fibers of π. This action induces a grading $\mathcal{O}_X(X) = \bigoplus_{i \geq 0} \mathfrak{A}_i$ where $\mathfrak{A}_0 = \mathcal{O}_B(B)$, $\mathfrak{A}_1 \neq \{0\}$, and $\mathfrak{A}_i$ consists of the $\mathbb{G}_m$-quasi-invariants of weight i:

$$h \in \mathfrak{A}_i \Leftrightarrow \lambda.h = \lambda^i \cdot h \;\; \forall \lambda \in \mathbb{G}_m .$$

If $u \in \mathfrak{A}_1$ then the restriction of u to a general fiber of π yields a coordinate on this fiber. It follows that $\psi = (\mathrm{id}_B, u): X \to B \times \mathbb{A}^1$ is a birational morphism over B, hence an affine modification (see [**47**, Thm. 1.1]). Since ψ is $\mathbb{G}_m$-equivariant its exceptional divisor E, its center C, and its divisor D are $\mathbb{G}_m$-invariant. Since u is a $\mathbb{G}_m$-quasi-invariant of weight 1 it vanishes along the zero section $Z \subset X$ with order 1. Thus, one has $u^{-1}(0) = Z \cup F_1 \cup \ldots \cup F_n$ where $F_i = \pi^{-1}(b_i)$, $b_i \in B$, $i = 1, \ldots, n$. Then

$$E = F_1 \cup \ldots \cup F_n, \quad C = \{b_1, \ldots, b_n\} \times \{0\}, \quad \text{and} \quad D = \{b_1, \ldots, b_n\} \times \mathbb{A}^1 \subset B \times \mathbb{A}^1 .$$

So, ψ consists in blowing up a subscheme with support C and deleting the proper transform of D. Therefore, ψ factorizes through the $\mathbb{G}_m$-equivariant fibered modification $\varrho_1: Z_1 \to B \times \mathbb{A}^1$ which consists in blowing up the reduced subscheme C and deleting the proper transform of D. Similarly, the resulting birational morphism of line bundles $X \to Z_1$ over B can be factorized over B into $X \to Z_2 \to Z_1$ where the center of $Z_2 \to Z_1$ is a reduced subscheme of C. After a finite number of steps we get a $\mathbb{G}_m$-equivariant resolution of indeterminacies of the inverse birational map $\psi^{-1}: B \times \mathbb{A}^1 \dashrightarrow X$, hence also a desired decomposition of ψ into a sequence (2.6) of fibered modifications.

(b) Under the assumptions of (b) consider the induced μ_d-action on $Z_0 = B \times \mathbb{A}^1$ identical on the second factor. In order that $\psi = (\mathrm{id}_B, u): X \to B \times \mathbb{A}^1$ were μ_d-equivariant it suffices to choose $u \in \mathfrak{A}_1^{\mu_d}$ being a μ_d-invariant. Since μ_d acts via automorphisms of the line bundle $\pi: X \to B$ it normalizes the $\mathbb{G}_m$-action on X. Hence it induced a representation of μ_d via automorphisms of the graded k-algebra $\mathcal{O}_X(X) = \bigoplus_{i \geq 0} \mathfrak{A}_i$. Let

$$\mathfrak{A}_1^{(i)} = \{\mathfrak{a} \in \mathfrak{A}_1 \,|\, \zeta.\mathfrak{a} = \zeta^i \mathfrak{a} \;\; \forall \zeta \in \mu_d\} .$$

Any element $\mathfrak{a} \in \mathfrak{A}_1$ belongs to the μ_d-invariant subspace E spanned by the orbit $\mu_d(\mathfrak{a})$. The finite dimensional representation of μ_d in E splits into a sum of one-dimensional representations. Consequently, $\mathfrak{a}$ can be written as a sum of μ_d-quasi-invariants. It follows that $\mathfrak{A}_1 = \bigoplus_{i=0}^{d-1} \mathfrak{A}_1^{(i)}$.

We claim that there exists a nonzero invariant $u \in \mathfrak{A}_1^{(0)} = \mathfrak{A}_1^{\mu_d}$. Indeed, for some $i \in \{0, \ldots, d-1\}$ there exists a nonzero μ_d-quasi-invariant $h \in \mathfrak{A}_1^{(i)}$ of weight i. The marking $z \in \mathfrak{A}_0 = \pi^*(\mathcal{O}_B(B))$ is a μ_d-quasi-invariant of weight 1, see Definition 2.1.1. Then $u = z^{d-i}h \in \mathfrak{A}_1^{\mu_d}$ as desired.

The resulting birational morphism $\psi = (\mathrm{id}_B, u): X \to B \times \mathbb{A}^1$ over B is μ_d-equivariant. So, this is an affine modification with μ_d-invariant center C and divisor D.

Hence ψ factorizes through the μ_d-equivariant fibered modification $\varrho_1: Z_1 \to B \times \mathbb{A}^1$ which consists in blowing up $B \times \mathbb{A}^1$ with center the reduced zero dimensional subscheme $C \subset B \times \{0\}$ and deleting the proper transform of D. By recursion one arrives at a sequence (2.6) of μ_d-equivariant morphisms. □

REMARK 2.4.5. In the notation as in the proof of Lemma 2.4.4(a) let $\operatorname{div} u = Z + \sum_{i=1}^n m_i F_i$. Then the effective divisor $\mathfrak{b} := m_1 b_1 + \ldots + m_n b_n \in \operatorname{Div}(B)$ in this proof can be replaced by any representative $\mathfrak{b}' \in |\mathfrak{b}|$. Indeed, let $\mathfrak{b}' = \mathfrak{b} + \operatorname{div} f$ for a rational function f on B. Then $u' = uf \in \mathfrak{A}_1 \subset \mathcal{O}_X(X)$ and $\operatorname{div} u' = Z + \pi^*(\mathfrak{b}')$.

Letting in (2.5) $G = \mu_d$ and extending this sequence on the right by those in (2.6) with a suitable new enumeration we arrive at our final sequence of fibered modifications.

COROLLARY 2.4.6. (a) *Any GDF surface $\pi: X \to B$ can be obtained starting with a product $X_0 = B \times \mathbb{A}^1$ via a sequence of fibered modifications*

$$X = X_m \xrightarrow{\varrho_m} X_{m-1} \longrightarrow \ldots \longrightarrow X_1 \xrightarrow{\varrho_1} X_0 = B \times \mathbb{A}^1 \tag{2.7}$$

such that the center of ϱ_i is contained in the exceptional divisor of ϱ_{i-1}.

(b) *Suppose furthermore that $\pi: X \to B$ is a marked GDF μ_d-surface. Then any intermediate surface X_i, $i = 0, \ldots, m-1$, comes equipped with the induced μ_d-action so that the morphisms $\varrho_{i+1}: X_{i+1} \to X_i$ and $\pi_i: X_i \to B$ are μ_d-equivariant.*

PROOF. Let us prove (a) leaving the proof of (b) to the reader. Both (2.5) and (2.6) are chosen well ordered, that is, the center of ϱ_i is contained in the exceptional divisor of ϱ_{i-1}. Let the centers of blowups in (2.5) and (2.6) are situated over the points $b_1, \ldots, b_n \in B$ and $b'_1, \ldots, b'_{n'} \in B$, respectively. Let $\mathfrak{b} = b_1 + \ldots + b_n \in \operatorname{Div}(B)$ and $\mathfrak{b}' = b'_1 + \ldots + b'_{n'} \in \operatorname{Div}(B)$. The linear system $|\mathfrak{b}'|$ is base point free. Hence $\mathfrak{b}' \in |\mathfrak{b}'|$ can be chosen so that $b_i \neq b'_j$ $\forall i, j$, see Remark 2.4.5. Then the blowups in (2.5) and (2.6) are independent (they commute, in a sense). So, we may perform the blowups in (2.5) and (2.6) at each step simultaneously so that the resulting sequence (2.7) will be well ordered. □

REMARKS 2.4.7. 1. The morphisms in (2.7) can be extended to suitable completions yielding a sequence of birational morphisms

$$\hat{X} = \hat{X}_m \xrightarrow{\hat{\varrho}_m} \hat{X}_{m-1} \longrightarrow \ldots \longrightarrow \hat{X}_1 \xrightarrow{\hat{\varrho}_1} \hat{X}_0 = \bar{B} \times \mathbb{P}^1, \tag{2.8}$$

where $\hat{\pi}_i: \hat{X}_i \to \bar{B}$ is a μ_d-equivariant $\mathbb{P}^1$-fibration extending $\pi_i: X_i \to B$ and $\hat{\varrho}_i: \hat{X}_i \to \hat{X}_{i-1}$ is a simultaneous contraction of a μ_d-invariant union of disjoint (-1)-components of $\hat{\pi}_i$-fibers, $i = 0, \ldots, m$. Inspecting the proof of Lemma 2.4.4 one can see that certain irreducible fibers of $\mathrm{pr}_1: \bar{B} \times \mathbb{P}^1 \to \bar{B}$ are replaced by chains of rational curves with sequences of weights of type $[[-1, -2, \ldots, -2, -1]]$, where the *weights* are understood as in Definition 2.2.2. This yields a completion $\hat{X}_m \to \bar{B}$ of $X_m \to B$ whose boundary $\hat{X}_m \setminus X_m$ is a simple normal crossings divisor. The section at infinity $\bar{B} \times \{\infty\}$ of $\mathrm{pr}_1: \bar{B} \times \mathbb{P}^1 \to \bar{B}$ gives rise to a section at infinity S of

$\hat{X} = \hat{X}_m \to \bar{B}$ with $S^2 = 0$. If the line bundle $\bar{X}_0 \to \bar{B}$ in (2.5) is nontrivial then the completion $(\hat{X}, \hat{D})$ is not pseudominimal.

2. It is easily seen that F has level l if and only if it appears for the first time on the surface X_l $(l \le m)$ in (2.7), see Definition 2.3.9. Thus, any fiber component F' of $\pi_l: X_l \to B$ has level $\le l$. If $\pi: X \to B$ is a marked GDF μ_d-surface then the level function is μ_d-invariant. Notice that the center of the blowup $\varrho_{l+1}: X_{l+1} \to X_l$ in (2.7) is situated on the union of the top level l fiber components in X_l.

DEFINITION 2.4.8 (*Trivializing completions*). The completion $(\hat{X}, \hat{D})$ of a GDF surface X fitting in (2.8) and the corresponding graph divisor $\mathcal{D}(\hat{\pi})$ will be called *trivializing.*

Notice that a trivializing completion is not necessarily pseudominimal, see Definition 2.2.1.

CHAPTER 3

Vector fields and natural coordinates

3.1. Locally nilpotent vertical vector fields

We say that a vector field v on a μ_d-variety *has weight* l if $\zeta.v = \zeta^l \cdot v\ \forall \zeta \in \mu_d$.

LEMMA 3.1.1. *Let* $\pi\colon X \to B$ *be a marked GDF* μ_d*-surface with a marking* $z \in \mathcal{O}_B(B) \setminus \{0\}$*. Then for any* $l = 0, \dots, m$ *the surface* X_l *in* (2.7) *admits a locally nilpotent regular* μ_d*-quasi-invariant vertical vector field* ∂_l *of weight* l *non-vanishing on the fiber components of the top level* l *and vanishing on the fiber components of* $z^{-1}(0)$ *of lower levels.*

PROOF. Consider the locally nilpotent vertical vector field $\partial_0 = \partial/\partial u$ on $X_0 = B \times \mathbb{A}^1$ where $\mathbb{A}^1 = \operatorname{spec} \mathbb{k}[u]$. Clearly, ∂_0 is invariant under the μ_d-action on $B \times \mathbb{A}^1$ identical on the second factor. The μ_d-equivariant fibered modification $\varrho_1\colon X_1 \to B \times \mathbb{A}^1$ over B transforms ∂_0 into a μ_d-invariant rational vertical vector field on X_1 with pole of order 1 along the fiber components of level 1. By induction, ∂_0 lifts to a μ_d-invariant rational vertical vector field on X_l with pole of order s on any fiber component of level s where $1 \le s \le l$ and no other pole.

Since the marking $z \in \mathcal{O}_B(B) \setminus \{0\}$ is μ_d-quasi-invariant of weight 1 then $\partial_l := z^l \partial/\partial u$ generates a regular locally nilpotent μ_d-quasi-invariant vertical vector field on X_l of weight l non-vanishing on the fiber components of level l and vanishing on the fiber components of smaller levels. □

3.2. Standard affine charts

NOTATION 3.2.1. Let $\pi\colon X \to B$ be a marked GDF μ_d-surface with a marking $z \in \mathcal{O}_B(B)$ where $z^*(0) = b_1 + \ldots + b_n$. For any $i = 1, \dots, n$ consider in B the affine chart $B_i = B \setminus \{b_1, \dots, b_{i-1}, b_{i+1}, \dots, b_n\}$ about the point b_i. So, $\operatorname{div}(z|_{B_i}) = b_i$.

Given a surface X_l from (2.7) we let $F_{i,1}, \dots, F_{i,n_i}$ be the components of the fiber $\pi_l^{-1}(b_i)$. Consider the $\mathbb{G}_a$-action H_l on X_l along the fibers of π_l generated by the locally nilpotent vector field ∂_l as in Lemma 3.1.1.

PROPOSITION 3.2.2. *In the notation as above the following hold.*

- *For any* $i \in \{1, \dots, n\}$ *and* $j \in \{1, \dots, n_i\}$ *there is a unique* standard *affine chart* $U_{i,j} \supset F_{i,j}$ *in* X_l *such that* $U_{i,j} \cong_{B_i} B_i \times \mathbb{A}^1$*. These standard affine charts* $(U_{i,j})_{i,j}$ *form a covering of* X_l*;*
- *one has* $U_{i,j} \cap U_{i,j'} = U_{i,j} \setminus F_{i,j} = U_{i,j'} \setminus F_{i,j'}$ *for any* $1 \le j, j' \le n_i$*;*
- *the* μ_d*-action on* X_l *induces a* μ_d*-action by permutations on the collection* $(U_{i,j})$*;*
- $U_{i,j}$ *is invariant under any action of a connected algebraic group on* $X_l \to B$ *identical on* B*;*
- *for any* $t \le l$ *and any fiber component* $F_{i,j}$ *on level* t *the* H_t*-action is well defined and free on* $U_{i,j}$*;*

- *for any* $l, t \in \mathbb{Z}$ *with* $0 \le t < l \le m$ *the composition* $\varrho_{l,t} = \varrho_{t+1} \circ \ldots \circ \varrho_l : X_l \to X_t$ *sends a standard affine chart* $U_{i,j} \subset X_l$ *on level* t *isomorphically over* B_i *to a standard affine chart in* X_t.

PROOF. The assertions are evidently true for the product $X_0 = B \times \mathbb{A}^1$ in (2.7) with $n_i = 1\ \forall i$ and $U_{i,1}^{(0)} = \pi_0^{-1}(B_i) = B_i \times \mathbb{A}^1$. Suppose by recursion that they hold for a surface X_{l-1} in (2.7) and the collection $(U_{i,j}^{(l-1)})$ of standard affine charts on X_{l-1}. The μ_d-equivariant fibered modification $\varrho_l : X_l \to X_{l-1}$ in (2.7) consists in blowing up with center at a union of μ_d-orbits situated on special fiber components of the top level $l-1$ in X_{l-1} and deleting the proper transforms of these fiber components, see Remark 2.4.7.2. Let $F = F_{i,j}$ be one of these components, and let $U_F = U_{i,j}^{(l-1)}$ be the corresponding standard affine chart in X_{l-1}. Then the modification ϱ_l replaces F with new components, say, $F_1, \ldots, F_M$ of level l on X_l. The induced fibered modification of $U_F \cong_{B_i} B_i \times \mathbb{A}^1$ results in a GDF surface over B_i with the only degenerate fiber $\pi_l^{-1}(b_i) = \sum_{j=1}^M F_j$. Blowing up just one point, say, x_j one replaces F by F_j. Choosing local coordinates (z, u) in $U_F \cong_{B_i} B_i \times \mathbb{A}^1$ so that $u(x_j) = 0$, $z(x_j) = 0$ the latter affine modification consists in passing from $\mathcal{O}_{B_i}(B_i)[u]$ to $\mathcal{O}_{B_i}(B_i)[u/z] = \mathcal{O}_{B_i}(B_i)[u']$ where $u' = u/z$. This results in a standard affine chart $U_{F,j}^{(l)} \cong_{B_i} B_i \times \mathbb{A}^1$ in X_l. In total one obtains M such affine charts on X_l over U_F with the intersections as needed. For a fiber component F on X_{l-1} which does not contain any center of the modification ϱ_l we let $U_F^{(l)} = \varrho_l^{-1}(U_F^{(l-1)})$. It is easily seen that the desired conclusions hold for the resulting collection $(U_{F,j}^{(l)})$ of standard affine charts on X_l. We leave the details to the reader. □

3.3. Natural coordinates

Let $\pi : X \to B$ be again a marked GDF μ_d-surface with a marking $z \in \mathcal{O}_B(B)$ where $z^*(0) = b_1 + \ldots + b_n$.

DEFINITION 3.3.1 (*Natural local coordinates*). Fix a component F of a fiber $\pi_l^{-1}(b_i)$ on X_l, and let U_F be the standard affine chart in X_l about F. An isomorphism $U_F \cong_{B_i} B_i \times \mathbb{A}^1$ provides sections of $\pi_l|_{U_F} : U_F \to B_i$. Fixing such a section and using the vertical free $\mathbb{G}_a$-action on U_F one obtains a $\mathbb{G}_a$-equivariant isomorphism $U_F \cong_{B_i} B_i \times \mathbb{A}^1$ where $\mathbb{G}_a$ acts on the direct product via translations along the second factor. Fixing a coordinate u in $\mathbb{A}^1$ one gets a coordinate, say, $u = u_F$ in U_F.

The restriction $z|_{U_F}$ vanishes to order 1 along F and has no further zero. Hence (z, u_F) yields local coordinates in U_F near F. We call these *natural coordinates.* The local coordinates (z, u_F, v) in the standard affine chart $U_F \times \mathbb{A}^1$ in the cylinder $\mathcal{X}_l$ about the affine plane $\mathcal{F} = F \times \mathbb{A}^1 = \operatorname{Spec} \Bbbk[u_F, v]$ are also called *natural.*

LEMMA 3.3.2. *Let* $\pi : X \to B$ *be a marked GDF* μ_d*-surface, and let* X_l *be one of the surfaces in* (2.7). *Let* $\mathfrak{F}_l$ *be the collection of the components of* $z^*(0)$ *in* X_l *on the top level* l. *Then there exists a collection* $(z, u_F)_{F \in \mathfrak{F}_l}$ *of natural local coordinates such that* $(u_F)_{F \in \mathfrak{F}_l}$ *is a quasi-invariant of* μ_d *of weight* $-l$.

PROOF. For $F \in \mathfrak{F}_l$ let $\mu_e \subset \mu_d$ (where $e|d$) be the isotropy subgroup of F. The μ_e-action on U_F induces a μ_e-action on $B_i \times \mathbb{A}^1 \cong_{B_i} U_F$ (see Notation 3.2.1). The latter isomorphism yields a bijection between the sections of $\pi_l|_{U_F} : U_F \to B_i$ and the functions in $\mathcal{O}_{B_i}(B_i)$. Choosing an arbitrary such section and averaging over its μ_e-orbit one obtains a μ_e-invariant section, say, Z_F. Then Z_F can be taken for the zero locus of a coordinate function u_F in U_F. Let ∂_l be the vertical μ_d-quasi-invariant vector field on X_l of weight l constructed in Lemma 3.1.1. The coordinate function $u_F \in \mathcal{O}_{U_F}(U_F)$ with $u_F|_{Z_F} = 0$ and $\partial_l(u_F) = 1$ is unique. For $\zeta \in \mu_e$ the ratio $\zeta.u_F/u_F$ does not vanish, hence it is constant along any π_l-fiber. Thus, one has $\zeta.u_F = (\pi_l^* f) \cdot u_F$ for some $f \in \mathcal{O}^\times_{B_i}(B_i)$. From the relations $\partial_l \circ \zeta = \zeta^l \partial_l$ and $\partial_l(\pi_l^* f) = 0$ one deduces that $f = \zeta^{-l}$ is a constant, and so, u_F is a μ_e-quasi-invariant of weight $-l$.

For any component F' in the μ_d-orbit of F define natural coordinates $(z, u_{F'})$ in $U_{F'}$ in such a way that the collection of functions $(u_{F'})$ becomes μ_d-quasi-invariant of weight $-l$. Choosing a representative of any μ_d-orbit on $\mathfrak{F}_l$ and repeating the same procedure gives the desired collection of local coordinates. □

REMARKS 3.3.3. 1. If F is a component of $z^*(0)$ on X_l on level $l' < l$ then the μ_d-equivariant morphism $\varrho_l : X_l \to X_{l-1}$ restricts to an isomorphism on U_F. Hence one can define local coordinates (z, u_F) in U_F where F runs over all components of $z^*(0)$ in X_l in such a way that for a given $l' \le l$ the collection $(u_F)_{F \in \mathfrak{F}_{l'}}$ is a μ_e-quasi-invariant of weight $-l'$.

In the local coordinates (z, u_F) in $U_F \subset X_l$ of level $l' \le l$ the vertical vector field ∂_l on X_l constructed in Lemma 3.1.1 coincides with $z^{l-l'} \partial/\partial u_F$. In particular, in a top level chart U_F one has $\partial_l|_U = \partial/\partial u_F$.

2. In the case $e = 1$ our choice of a μ_e-invariant section is arbitrary, and the coordinate u_F in the standard affine chart U_F is defined up to a factor which is an invertible function lifted from the base and a shift along the u-axis. Hence for F of the top level one may consider that u_F does not vanish in any center of the blowup $\varrho_{l+1} : X_{l+1} \to X_l$ contained in F.

3.4. Special μ_d-quasi-invariants

In the sequel we need μ_d-invariant locally nilpotent derivations on the cylinders over GDF μ_d-surfaces. To this end we construct on such surfaces quasi-invariant functions of prescribed weights, see Corollary 3.4.2 below. Let us start with the following fact (cf. [**46**, Lem. 2.12]).

LEMMA 3.4.1. *Suppose we are given a finite group G, a character $\lambda \in G^\vee$, an affine G-variety Y, and a G-invariant closed subscheme Z of Y which is not necessarily reduced. Let $f \in \mathcal{O}_Z(Z)$ belongs to λ, that is, $f \circ g = \lambda(g) \cdot f \;\forall g \in G$. Then f admits a regular G-quasi-invariant extension to Y which belongs to λ.*

PROOF. Letting $A = \mathcal{O}_Y(Y)$ and $B = \mathcal{O}_Z(Z)$ the G-action yields graded decompositions $A = \bigoplus_{\chi \in G^\vee} A_\chi$ and $B = \bigoplus_{\chi \in G^\vee} B_\chi$. The piece A_χ (B_χ, respectively) consists of the G-quasi-invariants in A (in B, respectively) which belong to the character χ. The closed embedding $Z \hookrightarrow Y$ induces a surjection $\varphi : A \to B$ ([**43**, Thm. III.3.7]). We claim that φ restricts to a surjection $\varphi|_{A_\lambda} : A_\lambda \to B_\lambda$ for any $\lambda \in G^\vee$. Indeed, any $f \in B_\lambda$ admits an extension to a regular function $\tilde{f} \in A$ such that $\varphi(\tilde{f}) = f$. There is a unique decomposition $\tilde{f} = \sum_{\chi \in G^\vee} \tilde{f}_\chi$. Hence $f = \sum_{\chi \in G^\vee} \varphi(\tilde{f}_\chi)$. Since $f \in B_\lambda$ the summands $\varphi(\tilde{f}_\chi)$ with $\chi \ne \lambda$ vanish, and so, $f = \varphi(\tilde{f}_\lambda)$. Hence $\tilde{f}_\lambda \in A_\lambda$ is a desired G-quasi-invariant extension of f which belongs to λ. □

COROLLARY 3.4.2. *Let $\pi : X \to B$ be a marked GDF μ_d-surface, let X_l be one of the surfaces in (2.7), and let $\mathfrak{F}'_l$ be a μ_d-invariant subset of the set $\mathfrak{F}_l$ of top level* *fiber components in* X_l. *Consider a* μ_d*-quasi-invariant collection of natural local coordinates* $(z, u_F)_{F\in\mathfrak{F}_l}$ *as in Lemma 3.3.2. Then for any* $s \gg 1$ *one can find a* μ_d*-quasi-invariant function* $\tilde{u} \in \mathcal{O}_{X_l}(X_l)$ *of weight* $-l$ *such that*

(i) $\tilde{u} \equiv u_{F'} \bmod z^s$ *near* F' *if* $F' \in \mathfrak{F}'_l$ *and*
(ii) $\tilde{u} \equiv 0 \bmod z^s$ *near* F' *otherwise.*

PROOF. It suffices to apply Lemma 3.4.1 with $Y = X_l$, $G = \mu_d$, $Z = (z^s)^*(0)$ being the sth infinitesimal neighborhood of the union of the special fiber components in X_l, $\lambda(\zeta) = \zeta^{-l}$ for $\zeta \in \mu_d$, and the function $f \in \mathcal{O}_Z(Z)$ defined in the affine charts $Z \cap U_{F'}$ via $f|_{Z\cap U_{F'}} = u_{F'}|_{Z\cap U_{F'}}$ for $F' \in \mathfrak{F}'_l$ and $f|_{Z\cap U_{F'}} = 0$ otherwise. □

3.5. Examples of GDF surfaces of Danielewski type

We start with the classical Danielewski example.

EXAMPLE 3.5.1 (*Danielewski surfaces*). The Danielewski surface X_1 results from the affine modification $\varrho_1: X_1 \to X_0$ of the affine plane $X_0 = \mathbb{A}^2 = \operatorname{Spec} \Bbbk[z,u]$ with the divisor $z = 0$ and the center $I = (z, u^2 - 1)$. This consists in blowing up the points $x_1 = (0,1)$ and $x_{-1} = (0,-1)$ in $\mathbb{A}^2$ and deleting the proper transform of the affine line $z = 0$. Letting $A_0 = \mathcal{O}_{X_0}(X_0) = \Bbbk[z,u]$ and $A_1 = \mathcal{O}_{X_1}(X_1)$ one has

$$A_1 = A_0[(u^2-1)/z] = \Bbbk[z,u,t_1]/(zt_1 - u^2 + 1)\,.$$

The projections $\pi_0: X_0 \to B = \operatorname{Spec} \Bbbk[z]$ and $\pi_1: X_1 \to B$ are induced by the inclusions $\Bbbk[z] \hookrightarrow \Bbbk[z,u] \hookrightarrow \Bbbk[z,u,(u^2-1)/z]$. Thus, X_1 is given in $\mathbb{A}^3$ with coordinates (z,u,t_1) by equation

$$zt_1 - u^2 + 1 = 0\,.$$

The unique reducible fiber $\pi_1^*(0)$ of the GDF surface $\pi_1 = z|_{X_1}: X_1 \to B = \mathbb{A}^1$ consists of two disjoint affine lines (components of level one)

$$F_1 = \{z = 0, u = 1\} \quad \text{and} \quad F_{-1} = \{z = 0, u = -1\}\,.$$

The complement $X_1 \smallsetminus F_j$ for $j \neq i$ gives a standard affine chart $U_i \cong \mathbb{A}^2$ about F_i. The chart U_1 can be obtained via the affine modification of X_1 along divisor $z^*(0) = F_1 + F_{-1}$ with center the ideal $\mathbb{V}(F_1) = (z, u-1)$. Thus,

$$\mathcal{O}_{U_1}(U_1) = A_1\left[(u-1)/z\right] = \Bbbk[z,u_1] \quad \text{where} \quad u_1 = (u-1)/z = t_1/(u+1)\,.$$

Similarly, the standard affine chart on X_1 about F_{-1} is

$$U_{-1} = X_1 \smallsetminus F_1 = \operatorname{Spec} A_1[(u+1)/z] = \operatorname{Spec} \Bbbk[z,u_{-1}] \cong \mathbb{A}^2$$

where z and $u_{-1} = (u+1)/z = t_1/(u-1)$ are natural coordinates in U_{-1}. The locally nilpotent vertical vector field

$$\partial_1 = z\partial/\partial u + 2u\partial/\partial t_1$$

on X_1 restricts to $\partial/\partial u_i$ in U_i, $i = 1, -1$. The phase flow of ∂_1 yields a free $\mathbb{G}_a$-action on X_1. It is sent under ϱ_1 to the field $d\varrho_1(\partial_1) = z\partial_0 = z\partial/\partial u$ on X_0.

The second Danielewski surface X_2 is obtained via the affine modification $\varrho_2: X_2 \to X_1$ with the divisor $z^*(0)$ on X_1 and the center $I = (z, t_1) \subset A_1$. Thus, ϱ_2 consists in blowing up X_1 at the points $x_{1.i} = (0, i, 0) \in F_i$, $i = 1, -1$ (the origins of the affine planes $U_i \cong \mathbb{A}^2$, $i = 1, -1$) and deleting the proper transforms of the fiber components F_1 and F_{-1}. Letting $t_2 = t_1/z$ one obtains

$$A_2 := \mathcal{O}_{X_2}(X_2) = A_1[t_1/z] = \Bbbk[z,u,t_2]/(z^2 t_2 - u^2 + 1)\,.$$

Once again, X_2 is a GDF surface with a unique reducible fiber $z^*(0)$ consisting of two components of level 2. Iterating this procedure we arrive at a sequence of Danielewski surfaces

$$X_m = \operatorname{Spec} \Bbbk[z, u, t_m]/(z^m t_m - u^2 + 1), \quad m = 1, 2, \dots$$

along with a sequence of affine modifications fitting in (2.7)

$$\varrho_m : X_m \to X_{m-1}, \quad (z, u, t_m) \mapsto (z, u, t_{m-1}) \quad \text{where} \quad t_{m-1} = z t_m .$$

The only special fiber $z^*(0)$ in X_m is reduced and consists of two components of level m. The vector field $z^m \partial/\partial u$ on X_0 lifts to the locally nilpotent vertical vector field on X_m,

$$\partial_m = z^m \partial/\partial u + 2u\partial/\partial t_m .$$

Its phase flow defines a free $\mathbb{G}_a$-action on X_m. The latter action restricts in a standard affine chart on X_m to the standard $\mathbb{G}_a$-action via shifts in the vertical direction.

The extended divisor $D_{\mathrm{ext,m}}$ of a minimal completion $\bar{\pi}: \bar{X}_m \to \mathbb{P}^1$ has dual graph
(3.1)

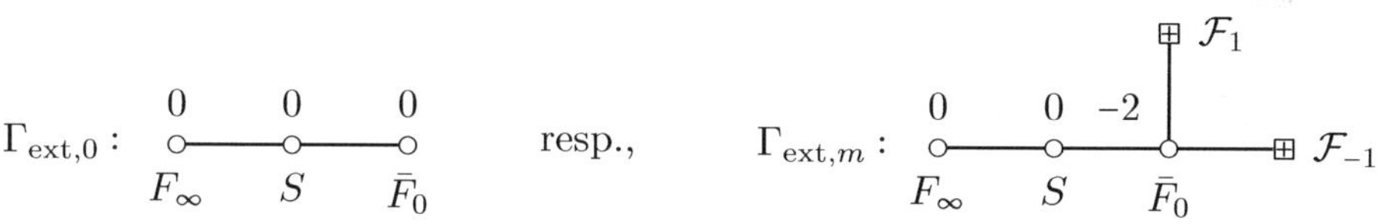

where for $m = 0$ one has $\bar{X}_0 \cong \mathbb{P}^1 \times \mathbb{P}^1$ and the distinguished (reduced and irreducible) fiber $\bar{F}_0$ is chosen just in sake of homogeneity of notation. For $m \geq 1$ the boxes $\mathcal{F}_1$ and $\mathcal{F}_{-1}$ stand for the chains $[[-2, \dots, -2, -1]]$ of length m so that $\mathcal{F}_i$ ends with the (-1)-feather $\bar{F}_i$ on level m, $i = 1, -1$ (cf. Example 7.1.3).

EXAMPLE 3.5.2. As an immediate generalization of the preceding example consider the surface X_m in $\mathbb{A}^3$ with equation $z^m t_m - q(u) = 0$ where $q \in \Bbbk[u]$ is a polynomial of degree $d \geq 2$ with simple roots. This is a GDF surface with projection $\pi = z|_{X_m}: X_m \to \mathbb{A}^1$. Letting $X_0 = \mathbb{A}^2$ one has a sequence of affine modifications (2.7) where $\varrho_i: X_i \to X_{i-1}$, $(z, u, t_i) \mapsto (z, u, t_{i-1} = z t_i)$. The vector field $z^m \partial/\partial u$ on $X_0 = \mathbb{A}^2$ lifts to the locally nilpotent vertical vector field on X_m,

$$\partial_m = z^m \partial/\partial u + q'(u)\partial/\partial t_m$$

which generates a free vertical $\mathbb{G}_a$-action on X_m. The dual graph $\Gamma_{\mathrm{ext,m}}$ of the pseudominimal completion $\bar{\pi}: \bar{X}_m \to \mathbb{P}^1$ differs from the graph in diagram (3.1) in one aspect: instead of two contractible chains $\mathcal{F}_1$ and $\mathcal{F}_{-1}$, it has d such chains $\mathcal{F}_j$, $j = 1, \dots, d$ of the same length m which are the branches of the fiber tree $\Gamma_0(\bar{\pi})$ in the root $\bar{F}_0$.

EXAMPLE 3.5.3 (*GDF surfaces given by equations*). In $\mathbb{A}^3$ with coordinates (z, u, t_1) consider the surface $X_1 = \{z t_1 - q_1(u) = 0\}$ where $q = q_1 \in \Bbbk[u]$ is a polynomial of degree $d \geq 1$ with simple roots $\alpha_1, \dots, \alpha_d$. The projection $\pi_1 = z|_{X_1}: X_1 \to \mathbb{A}^1$ makes of X_1 a GDF surface with a unique reducible fiber $\pi_1^{-1}(0)$ consisting of d components $F_1, \dots, F_d$ on level 1, see Example 3.5.2 with $m = 1$. The projection $(z, u, t_1) \mapsto (z, u)$ represents X_1 as a result of the fibered modification $\varrho_1: X_1 \to X_0 = \mathbb{A}^2$ which contracts F_i to the point $P_i = (0, \alpha_i) \in X_0$, $i = 1, \dots, d$.

Let further $q_2 \in \Bbbk[u,t_1]$ be such that, for each $i = 1,\dots,d$, either $q_2(\alpha_i,t_1) \in \Bbbk[t_1]$ has $m_i = \deg q_2(\alpha_i,t_1) > 0$ simple roots $\beta_{i,1},\dots,\beta_{i,m_i}$, or $q_2(\alpha_i,t_1) = 0$; in the latter case we let $m_i = 0$. Consider the complete intersection $V_2 \subset \mathbb{A}^4$ given in coordinates (z,u,t_1,t_2) by

$$zt_1 - q_1(u) = 0, \quad zt_2 - q_2(u,t_1) = 0\,.$$

There is a unique irreducible component X_2 of V_2 which dominates the z-axis, while the other components are disjoint affine planes contained in the hyperplane $z = 0$. Let $P_{i,j} = (0,\alpha_i,\beta_{i,j}) \in F_i$. The projection $(z,u,t_1,t_2) \mapsto (z,u,t_1)$ defines a fibered modification $\sigma_2\colon X_2 \to X_1$ along divisor $z^*(0) = \sum_{i=1}^d F_i$ with a reduced center

$$\bigcup_{m_i=0} F_i \cup \bigcup_{m_i>0} (P_{i,1}+\ldots+P_{i,m_i})\,.$$

The projection $\pi_2 = z|_{X_2}\colon X_2 \to \mathbb{A}^1$ makes of X_2 a GDF surface with a unique reduced fiber over $z=0$. One has $X_2 \setminus \pi_2^{-1}(0) \cong_{\mathbb{A}^1_*} \mathbb{A}^1_* \times \mathbb{A}^1$, that is, $\pi_2\colon X_2 \to \mathbb{A}^1$ is a Danielewski-Fieseler surface as defined in [**19**]. The fiber $\pi_2^{-1}(0) \subset X_2$ has $d - c_2$ components F_i of level 1 and $c_2 = m_1 + \dots + m_d$ components $F_{i,j}$ of level 2.

The graph $\Gamma_0(\pi_2)$ is a rooted tree with a root $\bar F_0$ of level 0, d vertices $\bar F_1,\dots,\bar F_d$ on level 1, and c_2 vertices $\bar F_{i,j}$, $i = 1,\dots,d, j = 1,\dots,m_i > 0$ on level 2 where $\bar F_{i,j}$ is a neighbor of $\bar F_i$. Clearly, any rooted tree Γ of height 3 can be realized as $\Gamma_0(\pi_2)$. Moreover, any Danielewski-Fieseler surface $\pi\colon X \to \mathbb{A}^1$ with $\Gamma_0(\pi) \cong \Gamma$ can be obtained in this way.

The vector field $z^2\partial/\partial u$ on $X_0 = \mathbb{A}^2$ lifted to X_2 extends in $\mathbb{A}^4$ to a locally nilpotent vector field

$$\partial_2 = z^2\frac{\partial}{\partial u} + zq_1'(u)\frac{\partial}{\partial t_1} + \left(z\frac{\partial q_2}{\partial u}(u,t_1) + q_1(u)\frac{\partial q_2}{\partial t_1}(u,t_1)\right)\frac{\partial}{\partial t_2}\,.$$

The associated vertical $\mathbb{G}_a$-action on X_2 is identical on the components F_i of level 1 (which correspond to $m_i = 0$) and is free on the components $F_{i,j}$ of level 2 and in the complement $X_2 \setminus \pi_2^{-1}(0)$.

By a recursive procedure one can realize in this way any Danielewski-Fieseler surface (cf. [**20**]). For instance, in the case of the Danielewski surface X_m from Example 3.5.2 one arrives at a system

$$zt_1 - p(u) = 0, \quad zt_2 - t_1 = 0, \quad \dots, \quad zt_m - t_{m-1} = 0\,,$$

which reduces to a single equation $z^m t_m - p(z) = 0$ defining the original proper embedding $X_m \hookrightarrow \mathbb{A}^3$.

CHAPTER 4

Relative flexibility

4.1. Definitions and the main theorem

NOTATION 4.1.1. Let $\pi\colon X \to B$ be a GDF surface, and let $\mathcal{X} = X \times \mathbb{A}^1$ be the cylinder over X. We let

$$\mathrm{SAut}_B(\mathcal{X}) = \langle \exp(\partial) \mid \partial \in \mathrm{LND}(\mathcal{O}_{\mathcal{X}}(\mathcal{X})),\ \partial(z) = 0\rangle$$

be the subgroup of the group $\mathrm{Aut}(\mathcal{X})$ generated by the automorphisms of $\mathcal{X}$ over B that are exponentials of locally nilpotent derivations from $\mathrm{LND}(\mathcal{X})$ cf. Section 1.1. For a component $\mathcal{F} = F \times \mathbb{A}^1$ of $z^*(0)$ in $\mathcal{X}$ any $\varphi \in \mathrm{SAut}_B(\mathcal{X})$ stabilizes the standard affine chart $U_F \times \mathbb{A}^1 \subset \mathcal{X}$ about $\mathcal{F}$ with natural coordinates (z, u_F, v), see Proposition 3.2.2 and Definition 3.3.1. Furthermore, for any such $\mathcal{F}$ the restriction $\varphi|_{U_F\times\mathbb{A}^1}$ preserves the volume form $dz \wedge du_F \wedge dv$ on $U_F \times \mathbb{A}^1$, that is, the Jacobian determinant of $\varphi|_{U_F\times\mathbb{A}^1}$ equals 1, see [**2**, Lem. 4.10].

DEFINITION 4.1.2 (*Relative flexibility*). We say that the cylinder $\mathcal{X} = X \times \mathbb{A}^1$ is *relatively flexible* (RF, for short) if for any natural $s \geq 1$, any collection $\mathfrak{F}$ of top level components $\mathcal{F} = F \times \mathbb{A}^1$ of $z^*(0)$ in $\mathcal{X}$, and any collection of pairs of ordered finite subsets $\Sigma_{\mathcal{F}} = \{x_1, \dots, x_M\}$ and $\Sigma'_{\mathcal{F}} = \{x'_1, \dots, x'_M\}$ in $\mathcal{F}$ of the same cardinality $M = M(\mathcal{F}) > 0$ where $\mathcal{F}$ runs over $\mathfrak{F}$, there exists an automorphism $\varphi \in \mathrm{SAut}_B\,\mathcal{X}$ which satisfies the conditions

- (α) $\varphi(x_\nu) = x'_\nu$ with prescribed volume preserving s-jets at x_ν, $\nu = 1, \dots, M(\mathcal{F})$ provided these jets preserve locally the fibers of $\mathcal{X} \to B$;
- (β) $\varphi|_{U_F\times\mathbb{A}^1} \equiv \mathrm{id} \mod z^s$ near $\mathcal{F} = F \times \mathbb{A}^1$ for any $\mathcal{F} \notin \mathfrak{F}$.

We say that *the condition RF(l,s) holds for X* if the relative flexibility holds for the cylinder over the surface X_l in (2.7) for a given $s \geq 1$.

DEFINITION 4.1.3 (*Equivariant relative flexibility*). Let $\pi\colon X \to B$ be a marked GDF μ_d-surface. We let $\mathcal{X}(k)$ denote the cylinder $\mathcal{X} = X \times \mathbb{A}^1$ equipped with a product μ_d-action where μ_d acts on the second factor via $(\zeta, v) \mapsto \zeta^k v$ for all $v \in \mathbb{A}^1$ and $\zeta \in \mu_d$. Assume that the collection $\mathfrak{F}$ of fiber components as in Definition 4.1.2 along with the finite sets $\Sigma = \bigcup_{F\in\mathfrak{F}} \Sigma_{\mathcal{F}}$ and $\Sigma' = \bigcup_{F\in\mathfrak{F}} \Sigma'_{\mathcal{F}}$ are μ_d-invariant, and the correspondence $\Sigma_{\mathcal{F}} \mapsto \Sigma'_{\mathcal{F}}$ is μ_d-equivariant.

We say that $\mathcal{X}(k)$ is *μ_d-relatively flexible* if one can choose a μ_d-equivariant automorphism $\varphi \in \mathrm{SAut}_{\mu_d,B}\,\mathcal{X}$ as in Definition 4.1.2 provided that

- (α_1) the collection of prescribed s-jets is μ_d-invariant;
- (α_2) if the stabilizer $\mu_e \subset \mu_d$ of $\mathcal{F}$ is nontrivial and $x_\nu \in \Sigma_{\mathcal{F}}$ is an isolated fixed point of the μ_e-action on $\mathcal{F}$ then $x'_\nu = x_\nu$ and the prescribed s-jet at x_ν is the one of the identity.

If the cylinder $\mathcal{X}_l(k)$ satisfies the above conditions for a given $s > 1$ then we say that *the μ_d-equivariant condition RF(l,k,s) holds for X*.

The main result of this section is the following

THEOREM 4.1.4. *Consider a marked GDF μ_d-surface $\pi: X \to B$ along with a trivializing sequence* (2.7). *Then the μ_d-equivariant condition RF$(l,-l,s)$ holds for X with arbitrary $s \geq 1$ and $l \in \{0,\dots,m\}$.*

The proof is done in Section 4.3.

4.2. Transitive group actions on Veronese cones

Let us recall the notion of a saturated set of locally nilpotent derivations (see [**2**, Def. 2.1]). For a vector field ∂ on a variety X and an automorphism $g \in \operatorname{Aut} X$ we let $\operatorname{Ad}(g)(\partial) = dg(\partial) \circ g^{-1}$.

DEFINITION 4.2.1 (*Saturation*). Consider an affine variety $X = \operatorname{Spec} A$ over $\Bbbk$. A set $\mathcal{N}$ of locally nilpotent regular vector fields on X (that is, of locally nilpotent derivations of the affine $\Bbbk$-algebra $A = \mathcal{O}_X(X)$) is called *saturated* if

(i) for any $\partial \in \mathcal{N}$ and $a \in \ker \partial$ the *replica* $a\partial$ belongs to $\mathcal{N}$, and
(ii) $\operatorname{Ad}(g)(\partial) \in \mathcal{N}$ $\forall g \in G$, $\forall \partial \in \mathcal{N}$ where $G = \langle \exp \partial \,|\, \partial \in \mathcal{N} \rangle \subset \operatorname{Aut} A$.

LEMMA 4.2.2. *Given a set $\mathcal{N} \subset \operatorname{Der} A$ of locally nilpotent derivations satisfying* (i) *consider the group $G \subset \operatorname{Aut} A$ as in* (ii) *generated by* $\exp(\mathcal{N})$. *Then the set of locally nilpotent derivations*

$$\mathcal{N}_1 = \{\operatorname{Ad}(g)(\partial) \,|\, g \in G,\ \partial \in \mathcal{N}\}$$

is saturated and generates the same group G.

PROOF. It is not difficult to see that $\mathcal{N}_1$ satisfies (i). Let $G_1 = \langle \exp \partial \,|\, \partial \in \mathcal{N}_1 \rangle$ be the group generated by $\mathcal{N}_1$. We claim that $G_1 = G$, and so, (ii) follows by the chain rule. Indeed, an automorphism $g \in \operatorname{Aut} X$ sends a vector field ∂ on X into the vector field ∂' on X such that $\partial'(g(x)) = dg(\partial(x))$ $\forall x \in X$. Hence $\partial' = \operatorname{Ad}(g)(\partial)$. On the other hand, if ∂ is locally nilpotent with the phase flow $\exp(t\partial) \in \operatorname{Aut} X$, $t \in k$, then for the phase flow $\exp(t\partial') \in \operatorname{Aut} X$, $t \in k$, one has $\exp(t\partial') = g \circ \exp(t\partial) \circ g^{-1}$. Since $\exp(t\partial) \in G$ it follows that $\exp(t\partial') \in G$ for any $g \in G$ and $\partial \in \mathcal{N}$. Thus $\exp(t\partial') \in G$ for any $\partial' \in \mathcal{N}_1$, and so, $G_1 = G$, as claimed. □

4.2.3. Given $c, d \in \mathbb{N}$ consider the affine plane $\mathbb{A}^2 = \operatorname{Spec} \Bbbk[u,v]$ equipped with the μ_d-action

$$\zeta.(u,v) = (\zeta^{-c}u, \zeta^{-c}v) \quad \forall \zeta \in \mu_d\,. \tag{4.1}$$

This action is not effective, in general. However, it restricts to an effective action of the subgroup $\mu_e \subset \mu_d$ where $e = d/\gcd(c,d)$. The quotient $V_e = \mathbb{A}^2/\mu_e = \mathbb{A}^2/\mu_d$ is a Veronese cone.

Consider also the locally nilpotent vector fields $\sigma_1 = v\frac{\partial}{\partial u}$ and $\sigma_2 = u\frac{\partial}{\partial v}$ on $\mathbb{A}^2$. The associated one-parameter groups $(u,v) \mapsto (u+tv, v)$ and $(u,v) \mapsto (u, v+tu)$, $t \in k$, generate the standard $\mathbf{SL}_2$-action on $\mathbb{A}^2$. Notice that σ_1 and σ_2 are μ_d-invariant and the μ_d-action on $\mathbb{A}^2$ commutes with the $\mathbf{SL}_2$-action. Hence the $\mathbf{SL}_2$-action descends to the Veronese cone V_e.

NOTATION 4.2.4. Given $s \geq 2$ consider the μ_d-invariant replicas

$$\sigma_{1,f} = v^{ds} f(v^d)\sigma_1 \text{ of } \sigma_1 \quad \text{and} \quad \sigma_{2,g} = u^{ds} g(u^d)\sigma_2 \text{ of } \sigma_2 \quad \text{where} \quad f, g \in \Bbbk[t]\,.$$

These vector fields are chosen so that $\sigma_{1,f}$ vanishes modulo v^s on the axis $v = 0$, and $\sigma_{2,g}$ vanishes modulo u^s on the axis $u = 0$. Hence $\varphi_f := \exp(\sigma_{1,f})$ fixes the axis $v = 0$

pointwise along with its infinitesimal neighborhood of order s, and $\psi_g := \exp(\sigma_{2,g})$ does the same with the axis $u = 0$. The subgroup

$$G = \langle \varphi_f,\ \psi_g \,|\, f, g \in \Bbbk[t]\rangle \subset \mathrm{SAut}(\mathbb{A}^2) \tag{4.2}$$

acts identically on the infinitesimal neighborhood of order s of the origin, is transitive in $\mathbb{A}^2 \smallsetminus \{\bar{0}\}$, and commutes with the μ_d-action (4.1) on $\mathbb{A}^2$.

Consider the normal subgroup $H \lhd G$ of all automorphisms $\alpha \in G$ of the from

$$\alpha = \varphi_1 \cdot \psi_1 \cdot \ldots \cdot \varphi_\nu \cdot \psi_\nu \tag{4.3}$$

where $\varphi_i = \varphi_{f_i}$ and $\psi_i = \psi_{g_i}$, $i = 1, \ldots, \nu$ verifying the condition

$$\varphi_1 \cdot \varphi_2 \cdot \ldots \cdot \varphi_\nu = \mathrm{Id}\,. \tag{4.4}$$

PROPOSITION 4.2.5. *With the notation as before, let $(O_1, \ldots, O_M)$ and $(O_1', \ldots, O_M')$ be two collections of distinct μ_d-orbits in $\mathbb{A}^2$ with* $\operatorname{card} O_i = \operatorname{card} O_i'$ *for $i = 1, \ldots, M$. For every $i = 1, \ldots, M$ choose a representative $x_i \in O_i$. In the case* $e = 1$ *suppose in addition that the singletons O_i and O_i' are different from the origin. Then there exists an automorphism $\alpha \in H$ such that*

(i) *$\alpha(O_i) = O_i'$ for $i = 1, \ldots, M$ and*
(ii) *α has prescribed values of volume-preserving s-jets at the points x_i, $i = 1, \ldots, M$ provided that for* $e > 1$ *and $O_i = \{\bar{0}\}$ for some $i \in \{1, \ldots, M\}$ this prescribed s-jet at the origin is the s-jet of the identity.* [1]

PROOF. Consider the Veronese cone $V_e = \mathbb{A}^2/\mu_d = \operatorname{Spec} \Bbbk[u, v]^{\mu_d}$, and let $\varrho\colon \mathbb{A}^2 \to V_e$ be the quotient morphism. The cone V_e is smooth outside the vertex $\bar{0} \in V_e$. Since the G-action on $\mathbb{A}^2$ is transitive in $\mathbb{A}^2 \smallsetminus \{\bar{0}\}$ and commutes with the μ_d-action it descends to a G-action on V_e transitive in $V_e \smallsetminus \{\bar{0}\}$. The μ_d-invariant locally nilpotent vector fields $\sigma_{1,f}$ and $\sigma_{2,g}$ also descend to V_e. The set $\mathcal{N}$ of all these vector fields on V_e satisfies condition (i) of Definition 4.2.1. By Lemma 4.2.2 the group $G \subset \operatorname{Aut} V_e$ is generated as well by a saturated set $\mathcal{N}_1$ of locally nilpotent vector fields on the cone V_e. Therefore one can apply Theorems 2.2 and 4.14 from [**2**].

Suppose first that $\{\bar{0}\}$ is not among the O_i's. By [**2**, Thm. 2.2], G acts infinitely transitively in $V_e \smallsetminus \{\bar{0}\}$. It follows that there exists $\alpha \in G$ which sends the points $y_i = \varrho(O_i) \in V_e$ into the points $y_i' = \varrho(O_i')$, $i = 1, \ldots, M$. Acting in $\mathbb{A}^2$ this α transforms the orbit O_i into O_i' for every $i = 1, \ldots, M$. Thus α verifies (i).

By [**2**, Thm. 4.14] one can find $\alpha \in G$ verifying (i) with a prescribed volume-preserving s-jet at each point $y_i = \varrho(O_i) \in V_e$, $i = 1, \ldots, M$. Since ϱ is a local isomorphism near a chosen point $x_i \in O_i$ over y_i and near its image $\alpha(x_i) \in O_i'$ one may prescribe a volume-preserving s-jet of α at x_i with the given zero term $\alpha(x_i)$.

If $e \geq 2$ and, say, $O_1 = \{\bar{0}\}$ then also $O_1' = \{\bar{0}\}$. Indeed, any μ_e-orbit different from $\{\bar{0}\}$ contains $e > 1$ points. Since $\sigma_{1,f}, \sigma_{2,g} \equiv 0 \mod (u, v)^s$ for any $f, g \in \Bbbk[t]$, see Notation 4.2.4, one has $\alpha \equiv \mathrm{id} \mod (u, v)^s$ for any $\alpha \in G$. Thus automatically $\alpha(\bar{0}) = \bar{0}$ and, moreover, the s-jet at the origin of any $\alpha \in G$ is the one of the identity map.

In the case $e = 1$ one has $V_e = \mathbb{A}^2$ and every orbit O_i and O_i' is a singleton different from $\{\bar{0}\}$ by assumption. Then the argument in the proof works without change.

[1] Instead of prescribing the value of an s-jet in a single point of a μ_e-orbit one might prescribe a μ_e-invariant collection of s-jets at the points of the orbit.

It remains to find such an automorphism in the subgroup H. Due to the infinite transitivity of G in $V_e \setminus \{\bar{0}\}$ one can find $\beta \in G$ such that for every $i = 1, \ldots, M$ the image $\beta(\varrho(O_i'))$ is located in the line $v = 0$ in V_e. By the preceding there exists $\alpha \in G$ such that $\alpha \circ \beta(\varrho(O_i)) = \beta(\varrho(O_i'))$ for all $i = 1, \ldots, M$ where α has prescribed volume preserving s-jets at these points. Since β is volume-preserving (see [**2**, Lemma 4.10]) one can find $\alpha_1 = \beta^{-1} \circ \alpha \circ \beta \colon \varrho(O_i) \mapsto \varrho(O_i')$ with prescribed volume preserving r-jets at the points $y_i = \varrho(O_i)$, $i = 1, \ldots, M$.

Decomposing α as in (4.3) consider the automorphism $\varphi_0 = (\varphi_1 \cdot \ldots \cdot \varphi_\nu)^{-1} \in G$. Since $\varphi_0 \cdot \varphi_1 \cdot \ldots \cdot \varphi_\nu = \mathrm{id}$ replacing φ_1 by $\varphi_0 \circ \varphi_1$ one obtains an automorphism $\alpha' = \varphi_0 \cdot \alpha \in H$. The s-jet of α' at each point $\beta(\varrho(O_i))$ is the same as the one of α. Indeed, $\varphi_0(\beta(\varrho(O_i'))) = \beta(\varrho(O_i'))$ since $\beta(\varrho(O_i')) \subset \{v = 0\}$, and φ_0 is identical on the sth infinitesimal neighborhood of this line. Since the subgroup $H \lhd G$ is normal one has $\alpha_1' = \beta^{-1} \circ \alpha \circ \beta \in H$ where $\alpha_1' \colon \varrho(O_i) \mapsto \varrho(O_i')$ and α_1' has prescribed volume preserving s-jets at the points $\varrho(O_i)$, $i = 1, \ldots, M$. Thus, $\alpha_1' \in H$ satisfies both (i) and (ii). □

4.3. Relatively transitive group actions on cylinders

NOTATION 4.3.1. Let $\pi \colon X \to B$ be a marked GDF μ_d-surface, and let X_l be one of the surfaces in (2.7). We fix quasi-invariant natural coordinates in the standard affine charts U_F in X_l so that the conventions of Lemma 3.3.2 and Remark 3.3.3.2 are fulfilled.

Fix also a μ_d-invariant collection $\mathfrak{F}$ of top level fiber components of $z^{-1}(0)$ in X_l. For $F \in \mathfrak{F}$ let $\mathcal{F}_0$ be the μ_d-orbit of F in X_l. For $s \geq 2$ let $\tilde{u} = \tilde{u}(\mathcal{F}_0) \in \mathcal{O}_{X_l}(X_l)$ be a μ_d-quasi-invariant function of weight $-l$ which verifies conditions (i) and (ii) of Corollary 3.4.2. Let ∂_l be the μ_d-quasi-invariant vertical vector field of weight l on X_l as in Lemma 3.1.1. Given $f, g \in \Bbbk[t]$ consider the μ_d-invariant locally nilpotent derivations of the algebra $\mathcal{O}_{\mathcal{X}_l}(\mathcal{X}_l)$,

$$\tilde{\sigma}_{1,f} = v^{ds+1} f(v^d) \partial_l \quad \text{and} \quad \tilde{\sigma}_{2,g} = \tilde{u}^{ds+1} g(\tilde{u}^d) \partial/\partial v \,. \tag{4.5}$$

Letting F run over $\mathfrak{F}$, the corresponding automorphisms

$$\tilde{\varphi}_f = \exp(\tilde{\sigma}_{1,f}) \quad \text{and} \quad \tilde{\psi}_g = \exp(\tilde{\sigma}_{2,g})$$

in $\mathrm{SAut}_{\mu_d,B}(\mathcal{X}_l(-l))$ generate a subgroup

$$G_{\mathfrak{F}} = \langle \tilde{\varphi}_f,\ \tilde{\psi}_g \mid f, g \in \Bbbk[t] \rangle \subset \mathrm{SAut}_{\mu_d,B}(\mathcal{X}_l(-l)) \tag{4.6}$$

contained in the centralizer of the cyclic group induced by the μ_d-action on $\mathcal{X}_l(-l)$.

Consider further the normal subgroup $H_{\mathfrak{F}} \lhd G_{\mathfrak{F}}$,

$$H_{\mathfrak{F}} = \{\tilde{\alpha} = \tilde{\varphi}_1 \cdot \tilde{\psi}_1 \cdot \ldots \cdot \tilde{\varphi}_\nu \cdot \tilde{\psi}_\nu \in G_{\mathfrak{F}} \mid \tilde{\varphi}_1 \cdot \ldots \cdot \tilde{\varphi}_\nu = \mathrm{id}\} \,. \tag{4.7}$$

For $\tilde{u} = \tilde{u}(\mathcal{F}_0)$ one has $\tilde{u} \equiv 0 \mod z^s$ in $U_{F'} \times \mathbb{A}^1$ near $\mathcal{F}' = F' \times \mathbb{A}^1$ for any $F' \notin \mathfrak{F}$, see condition (ii) in Corollary 3.4.2. Hence $\tilde{\psi}_g \equiv \mathrm{id} \mod z^s$ in $U_{F'} \times \mathbb{A}^1$ near $\mathcal{F}'$ for any $F' \notin \mathfrak{F}$ and any $g \in \Bbbk[t]$. Due to (4.7) for any $\tilde{\alpha} \in H_{\mathfrak{F}}$ one has

$$\tilde{\alpha}|_{U_{F'} \times \mathbb{A}^1} = (\tilde{\varphi}_1 \cdot \tilde{\psi}_1 \cdot \ldots \cdot \tilde{\varphi}_\nu \cdot \tilde{\psi}_\nu)|_{U_{F'} \times \mathbb{A}^1} \equiv \mathrm{id} \mod z^s \quad \forall F' \notin \mathfrak{F} \,. \tag{4.8}$$

DEFINITION 4.3.2 (*s-reduced automorphism*). Let $F \subset X_l$ be a special fiber component and U_F be the standard affine chart about F in X_l. Consider the affine chart $U_F \times \mathbb{A}^1$ about the affine plane $\mathcal{F} = F \times \mathbb{A}^1 \simeq \mathbb{A}^2$ in the cylinder $\mathcal{X}_l(-l)$. The subgroup $G_{\mathfrak{F}} \subset \mathrm{SAut}_B\, \mathcal{X}_l(-l)$ preserves every fiber of the $\mathbb{A}^2$-fibration $\mathcal{X}_l(-l) \to B$ and, moreover, every fiber component. Hence any $\alpha \in G_{\mathfrak{F}}$ preserves the affine chart

$U_F \times \mathbb{A}^1$ (cf. Proposition 3.2.2). Choosing natural coordinates (z,u,v) in $U_F \times \mathbb{A}^1$ the restriction $\alpha|_{U_F\times\mathbb{A}^1}$ can be written as

$$\alpha|_{U_F\times\mathbb{A}^1}: (z,u,v) \mapsto \left(z, \sum_{i=0}^{\infty} z^i f_i(u,v), \sum_{i=0}^{\infty} z^i g_i(u,v)\right).$$

We say that α is *s-reduced* if $f_1 = \ldots = f_s = g_1 = \ldots = g_s = 0$, that is,

$$\alpha(z,u,v) \equiv (z, f_0(u,v), g_0(u,v)) \bmod z^s \tag{4.9}$$

in any such affine chart $U_F \times \mathbb{A}^1$ in $\mathcal{X}_l(-l)$.

LEMMA 4.3.3. (a) *A composition of s-reduced automorphisms is again s-reduced.*
(b) *Any automorphism $\tilde{\alpha} \in H_{\mathfrak{F}}$ is s-reduced.*

PROOF. The proof of (a) is straightforward. To prove (b), due to (a) it suffices to show that $\tilde{\varphi} = \tilde{\varphi}_f$ and $\tilde{\psi} = \tilde{\psi}_g$ are s-reduced for any $f,g \in \Bbbk[t]$. However, the latter is true only in the top level affine charts. Indeed, in a standard affine chart U_F on level $l' \le l$ in X_l one has $\partial_l|_{U_F} = z^{l-l'}\partial/\partial u$ (see Remark 3.3.3.1). Hence

$$\tilde{\varphi}|_{U_F\times\mathbb{A}^1} = \exp\left(v^{ds+1} f(v^d)\partial/\partial u\right): (z,u,v) \mapsto \left(z, u + z^{l-l'} v^{ds+1} f(v^d), v\right) \tag{4.10}$$

is s-reduced if $l' = l$, that is, in any top level affine chart.

Since $\tilde{u}|_{U_F} \equiv u \mod z^s$ if $l' = l$ and $\tilde{u}|_{U_F} \equiv 0 \mod z^s$ otherwise, near the affine plane $\mathcal{F} \subset U_F \times \mathbb{A}^1$ one has

$$\tilde{\psi}|_{U_F\times\mathbb{A}^1} = \exp\left(\tilde{u}^{ds+1} f(\tilde{u}^d)\partial/\partial v\right): (z,u,v) \mapsto \left(z, u, v + u^{ds+1} g(u^d)\right) \mod z^s \tag{4.11}$$

if $l' = l$ and $\tilde{\psi}|_{U_F\times\mathbb{A}^1} \equiv \mathrm{id} \mod z^s$ otherwise. In particular, any $\tilde{\psi} \in G_{\mathfrak{F}}$ is s-reduced. It follows that any automorphism

$$\tilde{\alpha} = \tilde{\varphi}_1 \cdot \tilde{\psi}_1 \cdot \ldots \cdot \tilde{\varphi}_\nu \cdot \tilde{\psi}_\nu \in G_{\mathfrak{F}}$$

is s-reduced in every top level affine chart $U_F \times \mathbb{A}^1$. If F has level $l' < l$ then $\tilde{\psi}_i|_{U_F\times\mathbb{A}^1} \equiv \mathrm{id} \mod z^s \ \forall i = 1,\ldots,\nu$. Hence for any $\tilde{\alpha} \in H_{\mathfrak{F}}$ one has by (4.8)

$$\tilde{\alpha}|_{U_F\times\mathbb{A}^1} = (\tilde{\varphi}_1 \cdot \ldots \cdot \tilde{\varphi}_\nu)|_{U_F\times\mathbb{A}^1} \equiv \mathrm{id} \mod z^s,$$

that is, $\tilde{\alpha}$ is s-reduced. □

PROPOSITION 4.3.4. *Let $\mathfrak{F}$ be a μ_d-invariant collection of top level components $\mathcal{F} = F \times \mathbb{A}^1 \subset \mathcal{X}_l(-l)$ of $z^*(0)$, and let $\Sigma, \Sigma' \subset \bigcup_{\mathcal{F}\in\mathfrak{F}} \mathcal{F}$ be two μ_d-invariant finite sets which meet every special fiber component $\mathcal{F} \in \mathfrak{F}$ with the same positive cardinality. Assume that for some $s > 0$ at each point $x \in \Sigma\cap\mathcal{F}$ we are given a volume preserving two-dimensional s-jet j_x of an automorphism $\mathcal{F} \to \mathcal{F}$ such that*

- *the zero term of j_x runs over Σ' when x runs over Σ;*
- *the collection $(j_x)_{x\in\Sigma}$ commutes with the μ_d-action on $\mathcal{X}_l(-l)$;*
- *if $e = e(\mathcal{F}) = d/\gcd(d,l) > 1$, see 4.2.3, and $x_0 \in \mathcal{F} \cap \Sigma$ is a fixed point of the stabilizer μ_e of $\mathcal{F}$ in μ_d then j_{x_0} is the s-jet of the identity.*

Then there exists a (μ_d-equivariant) automorphism $\tilde{\alpha} \in H_{\mathfrak{F}}$ such that

(i) $\tilde{\alpha}(\Sigma) = \Sigma'$;
(ii) *$\tilde{\alpha}$ has the prescribed two-dimensional s-jets at the points of Σ;*
(iii) $\tilde{\alpha}|_{U_F\times\mathbb{A}^1} \equiv \mathrm{id} \mod z^s \quad \forall \mathcal{F} \notin \mathfrak{F}$.

PROOF. Let $\mathcal{F} \in \mathfrak{F}$, and let $\mu_d(\mathcal{F})$ be the μ_d-orbit of $\mathcal{F}$ in $\mathcal{X}_l(-l)$. It suffices to construct such an automorphism $\tilde{\alpha} \in H_{\mathfrak{F}}$ assuming that $\mathfrak{F}$ consists of the components from the orbit $\mu_d(\mathcal{F})$. Indeed, then $\tilde{\alpha} \in H_{\mathfrak{F}}$ coincides with the identity modulo z^s near any special fiber component $\mathcal{F}' \notin \mathfrak{F}$. Composing such automorphisms $\tilde{\alpha}$ for different top level orbits one obtains a desired automorphism in the general case.

Furthermore, if (i) and (ii) hold for a special fiber component $\mathcal{F}$ then they automatically hold for any special fiber component $\mathcal{F}' \in \mu_d(\mathcal{F})$ due to the μ_d-invariance of the conditions and the μ_d-equivariance of the automorphisms $\tilde{\alpha} \in H_{\mathfrak{F}}$. Hence it suffices to take care of a particular $\mathcal{F} \in \mathfrak{F}$ equipped with two collections of orbits $\{O_i \cap \mathcal{F}\}_{i=1,\dots,\nu}$ and $\{O_i' \cap \mathcal{F}\}_{i=1,\dots,\nu}$ of the stabilizer μ_e of $\mathcal{F}$ in μ_d, see Proposition 4.2.5. Let $U_F \times \mathbb{A}^1$ be the standard affine chart about $\mathcal{F}$ equipped with μ_e-quasi-invariant natural local coordinates (z, u_F, v). Due to Remark 3.3.3.2 for $e = 1$ one may assume $O_i \neq \{\bar{0}\}$ $\forall i$ as needed in Proposition 4.2.5, see Notation 4.3.1.

Comparing (4.10) and (4.11) with (4.2) in Notation 4.2.4 one can see that the automorphisms $\tilde{\varphi}_f, \tilde{\psi}_g \in H_{\mathfrak{F}}$ restrict to

$$\tilde{\varphi}_f|_{\mathcal{F}} = \varphi_f \quad \text{and} \quad \tilde{\psi}_g|_{\mathcal{F}} = \psi_g\,,$$

respectively, where φ_f and ψ_g run over the generators of the subgroup $G \subset \mathrm{SAut}_{\mu_d}(\mathcal{F})$ when f, g run over $\Bbbk[t]$. Let $H \lhd G$ be as in 4.2.4. Applying Proposition 4.2.5 one can find an automorphism $\alpha = \varphi_1 \cdot \psi_1 \cdot \ldots \cdot \varphi_\nu \cdot \psi_\nu \in H$ satisfying in the affine plane $\mathcal{F} \cong \mathbb{A}^2$ conditions (i) and (ii) of this proposition. Extending every φ_i to $\tilde{\varphi}_i \in H_{\mathfrak{F}}$ and ψ_i to $\tilde{\psi}_i \in H_{\mathfrak{F}}$ one obtains an s-reduced automorphism $\tilde{\alpha} = \tilde{\varphi}_1 \cdot \tilde{\psi}_1 \cdot \ldots \cdot \tilde{\varphi}_\nu \cdot \tilde{\psi}_\nu \in H_{\mathfrak{F}}$, see Lemma 4.3.3(b). Since $\tilde{\alpha}$ also satisfies (4.8) in Notation 4.3.1 then (iii) holds, and so, $\tilde{\alpha}$ is a desired automorphism. □

PROOF OF THEOREM 4.1.4. Let $\pi: X \to B$ be a marked GDF μ_d-surface. We have to show that the μ_d-equivariant condition $\mathrm{RF}(l, -l, s)$ holds for X whatever are $s \geq 1$ and $l \in \{0, \dots, m\}$. It suffices to show that, given any μ_d-invariant collection $\mathfrak{F}$ of top level special fiber components in X_l and any two finite sets $\Sigma, \Sigma' \subset \bigcup_{F \in \mathfrak{F}} \mathcal{F}$ with the same μ_d-orbit structure and with $\operatorname{card}\Sigma_F = \operatorname{card}\Sigma'_F > 0$ $\forall F \in \mathfrak{F}$ where $\Sigma_F = \Sigma \cap \mathcal{F}$, there exists $\varphi \in \mathrm{SAut}_B\, \mathcal{X}_l(-l)$ such that the μ_d-equivariant versions of conditions (α) and (β) in Definition 4.1.2 are fulfilled.

By Proposition 4.3.4 one can find $\varphi \in H_{\mathfrak{F}} \subset \mathrm{SAut}_{\mu_d, B}\, \mathcal{X}_l(-l)$ verifying (i) and (ii) of Proposition 4.2.5 and condition (4.8). That is, φ is μ_d-equivariant, s-reduced, verifies (4.8), sends Σ to Σ', and has prescribed 2-dimensional r-jets (in the vertical planes) in the (chosen) points on each μ_d-orbit in Σ. Since φ is s-reduced it has as well the prescribed volume preserving three-dimensional s-jets in the given points. Hence φ satisfies conditions (α), (α_1), and (α_2) of Definitions 4.1.2 and 4.1.3. Due to (4.8), φ satisfies also condition (β) of Definition 4.1.2. □

4.4. A relative Abhyankar-Moh-Suzuki Theorem

We need in the sequel the following version of the Abhyankar-Moh-Suzuki Epimorphism Theorem.

PROPOSITION 4.4.1. *Let $\pi: X \to B$ be a GDF surface, let $\{F_1, \dots, F_t\}$ be a collection of top level special fiber components in X, and let $\mathcal{F}_i = F_i \times \mathbb{A}^1 \cong \mathbb{A}^2$, $i = 1, \dots, t$, be the corresponding components of the divisor $z^*(0)$ in the cylinder*

$\mathcal{X}$. *For every* $i = 1,\ldots,t$ *we fix in* $\mathcal{F}_i$ *a curve* $C_i \cong \mathbb{A}^1$. *Then there exists an automorphism* $\alpha \in \mathrm{SAut}_B(\mathcal{X})$ *such that* $\alpha(C_i) = F_i \times \{0\}$, $i = 1,\ldots,t$.

PROOF. Choose $i \in \{1,\ldots,t\}$, and let $F = F_i$, $\mathcal{F} = \mathcal{F}_i$, and $C = C_i \subset \mathcal{F}$. Our assertion follows by induction on i from the next claim.

CLAIM. There exists an automorphism $\beta = \beta_i \in \mathrm{SAut}_B(\mathcal{X})$ such that $\beta(C) = F \times \{0\}$ and $\beta(F' \times \{0\}) = F' \times \{0\}$ for any special fiber component $F' \neq F$.

Indeed, to deduce the assertion it suffices to apply this claim successively for $i = 1,\ldots,t$.

PROOF OF THE CLAIM. By Corollary 3.4.2 one can find $\tilde{u} \in \mathcal{O}(X)$ such that

(i) $\tilde{u}|_F = u_F$ where u_F is an affine coordinate on F;
(ii) $\tilde{u}|_{F'} = 0$ for any $F' \neq F$.

Consider the locally nilpotent derivations on $\mathcal{O}_{\mathcal{X}}(\mathcal{X})$,

$$\sigma_1 = \partial_l \quad \text{and} \quad \sigma_2 = \tilde{u}\frac{\partial}{\partial v}$$

where l is the highest level of the special fiber components of X and ∂_l is a vertical locally nilpotent vector field on X as in Lemma 3.1.1 so that $\partial_l(z) = 0$ and $\partial_l|_F = \partial/\partial u_F$. Consider the replicas

$$\sigma_{1,f} = f(v)\sigma_1 \quad \text{and} \quad \sigma_{2,g} = g(\tilde{u})\sigma_2 \quad \text{where} \quad f, g \in \Bbbk[t]\,.$$

Their exponentials

$$\varphi_f = \exp(\sigma_{1,f}), \quad \psi_g = \exp(\sigma_{2,g}) \in \mathrm{SAut}_B\mathcal{X}$$

generate a subgroup $\mathcal{H} \subset \mathrm{SAut}_B\,\mathcal{X}$. In the affine plane $\mathcal{F} \cong \mathrm{Spec}\,\Bbbk[u_F, v]$ one has

$$\varphi_f|_{\mathcal{F}}: (u_F, v) \mapsto (u_F + f(v), v) \quad \text{and} \quad \psi_g|_{\mathcal{F}}: (u_F, v) \mapsto (u_F, v + u_F g(u_F))\,.$$

In particular, $\mathcal{H}|_{\mathcal{F}}$ contains all the transvections, hence also the group $\mathbf{SL}(2,k)$. For $F' \neq F$ by virtue of (ii) the group $\mathcal{H}|_{\mathcal{F}'}$ is generated by the shears $\varphi_f|_{\mathcal{F}'}$. It follows that

- $\mathcal{H}|_{\mathcal{F}} = \mathrm{SAut}\,\mathcal{F} \cong \mathrm{SAut}\,\mathbb{A}^2$ and
- the coordinate line $F' \times \{0\} \subset \mathcal{F}'$ stays $\mathcal{H}$-invariant for any $F' \neq F$.

Now the claim follows by the Abhyankar-Moh-Suzuki Theorem. □

The next lemma allows to interchange the u- and v-axes in the top level special fiber components of $\mathcal{X} \to B$.

LEMMA 4.4.2. *Let* $\pi: X \to B$ *be a marked GDF* μ_d*-surface, let* X_l *be one of the surfaces in* (2.7), *and let* $\{(z, u_F)\}$ *be a quasi-invariant system of natural local coordinates in the standard local charts* U_F *about the special fiber components* F *in* X_l. *Given* $s > 1$ *there exists a* μ_d*-equivariant automorphism* $\tau \in \mathrm{SAut}_B\,\mathcal{X}_l(-l)$ *such that*

- $\tau|_{U_F \times \mathbb{A}^1}: (z, u_F, v) \mapsto (z, v, -u_F) \mod z^s$ *for any top level* F;
- $\tau|_{U_F \times \mathbb{A}^1} = \mathrm{id} \mod z^s$ *for any* F *of lower level.*

PROOF. Likewise in (4.5) we let

$$\tilde{\sigma}_1 = v\partial_l \quad \text{and} \quad \tilde{\sigma}_2 = -\tilde{u}\partial/\partial v \tag{4.12}$$

where ∂_l is the vertical vector field on X_l as in Lemma 3.1.1 and $\tilde{u} \in \mathcal{O}_{\mathcal{X}_l(-l)}(\mathcal{X}_l(-l))$ is a μ_d-quasi-invariant of weight $-l$ verifying conditions (i) and (ii) of Corollary 3.4.2. Letting $\tilde{\varphi} = \exp(\tilde{\sigma}_1)$ and $\tilde{\psi} = \exp(\tilde{\sigma}_2)$ by virtue of (i) and (ii) one obtains

$$\tilde{\varphi}|_{U_F \times \mathbb{A}^1} : (z, u_F, v) \mapsto (z, u_F + v, v) \mod z^s$$

and

$$\tilde{\psi}|_{U_F \times \mathbb{A}^1} : (z, u_F, v) \mapsto (z, u_F, v - u_F) \mod z^s$$

if F is of top level and $\tilde{\psi}|_{U_F \times \mathbb{A}^1} \equiv \mathrm{id} \mod z^s$ otherwise, cf. (4.10) and (4.11). Letting $\tau = \tilde{\varphi}\tilde{\psi}\tilde{\varphi}$ one gets

$$\tau|_{U_F \times \mathbb{A}^1} : (z, u_F, v) \mapsto (z, v, -u_F) \mod z^s$$

if F is of top level and $\tau|_{U_F \times \mathbb{A}^1} \equiv \mathrm{id} \mod z^s$ otherwise. □

We need as well the following versions of Lemma 4.4.2.

LEMMA 4.4.3. *Under the assumptions of Lemma 4.4.2 consider a μ_d-invariant subset $\Upsilon \subset \{b_1, \ldots, b_n\} = z^{-1}(0)$. Given a collection $\mathfrak{F}_\Upsilon(l)$ of top level special fiber components in $\pi_l^{-1}(\Upsilon) \subset X_l$ there exists a μ_d-equivariant automorphism $\tau \in \mathrm{SAut}_B\, \mathcal{X}_l(-l)$ such that*

$$\tau|_{U_F \times \mathbb{A}^1} : (z, u_F, v) \mapsto (z, v, -u_F) \mod z^s \tag{4.13}$$

if $F \in \mathfrak{F}_\Upsilon(l)$ and $\tau|_{U_F \times \mathbb{A}^1} \equiv \mathrm{id} \mod z^s$ otherwise.

PROOF. Choose a μ_d-invariant $h \in \mathcal{O}_B(B)$ such that $h - 1 \equiv 0 \mod z^s$ near each point $b_i \in \Upsilon$ and $h \equiv 0 \mod z^s$ near each point $b_i \notin \Upsilon$. Denote by the same letter the lift of h to $\mathcal{X}_l$. For the regular μ_d-quasi-invariant vector field ∂_l of weight l on $\mathcal{X}_l(-l)$ one has $\partial_l(h) = 0$ and $\partial h / \partial v = 0$. Let $\tilde{u} \in \mathcal{O}(X)$ be as in Corollary 3.4.2 and $\tilde{\sigma}_i$, $i = 1, 2$ be as in (4.12). Then the locally nilpotent vector fields

$$\tilde{\sigma}_{1,h} = h\tilde{\sigma}_1 \quad \text{and} \quad \tilde{\sigma}_{2,h} = h\tilde{\sigma}_2 \tag{4.14}$$

on $\mathcal{X}_l(-l)$ are μ_d-invariant. Using these derivations instead of $\tilde{\sigma}_i$, $i = 1, 2$ and proceeding as in the proof of Lemma 4.4.2 yields the result. □

CHAPTER 5

Rigidity of cylinders upon deformation of surfaces

5.1. Equivariant Asanuma modification

In the next lemma we introduce an equivariant version of the Asanuma modification. For the reader's convenience we repeat in (a) the statement of Lemma 1.4.5.

LEMMA 5.1.1. *Let $\pi: X \to B$ be a GDF surface, and let $\varrho: X' \to X$ be a fibered modification along a reduced principal divisor* $\operatorname{div} f$ *where $f \in \pi^*\mathcal{O}_B(B) \setminus \{0\}$ with center a reduced zero-dimensional subscheme $\mathbb{V}(I)$ where $I \subset \mathcal{O}_X(X)$ is an ideal, see Definition 2.4.1. Consider the principal divisor $\mathcal{D} = \mathbb{V}(f) \times \mathbb{A}^1$ on the cylinder $\mathcal{X} = X \times \mathbb{A}^1$ and the ideal $J = (I, v) \subset \mathcal{O}_{\mathcal{X}}(\mathcal{X})$ with support $\mathbb{V}(I) \times \{0\} \subset \mathbb{V}(f) \times \{0\}$. Then the following holds.*

(a) *The cylinder $\mathcal{X}' = X' \times \mathbb{A}^1$ is isomorphic to the affine modification Z of $\mathcal{X}$ along divisor $\mathcal{D}$ with center J. This isomorphism fits in the commutative diagram*

$$\begin{array}{ccccc} \mathcal{X}' & \xrightarrow{\cong} & Z & \longrightarrow & \mathcal{X} \\ & \searrow & \downarrow & \swarrow & \\ & & B & & \end{array} \tag{5.1}$$

where the vertical arrows are $\mathbb{A}^2$-fibrations over B.

(b) *Assume that the modification $\varrho: X' \to X$ is equivariant with respect to actions of a finite group G on X, X', and B. Assume also that the ideal I is G-invariant, the function f is G-quasi-invariant and belongs to a character $\chi \in G^\vee$. Define the G-action on the factor $\mathbb{A}^1$ of the cylinder $\mathcal{X} = X \times \mathbb{A}^1$ via the multiplication by a character $\lambda \in G^\vee$. Then the morphisms in (5.1) are G-equivariant where G acts on the factors $\mathbb{A}^1$ of the cylinder $\mathcal{X}' = X' \times \mathbb{A}^1$ via the multiplication by λ/χ. In particular, if $G = \mu_d$, $\chi: \zeta \mapsto \zeta^t$, and $\lambda: \zeta \mapsto \zeta^k$ then $\lambda/\chi: \zeta \mapsto \zeta^{k-t}$ $\forall \zeta \in \mu_d$.*

PROOF. For the proof of (a) see Lemma 1.4.5. Statement (b) follows since under its assumptions the variable v' in the proof of Lemma 1.4.5 belongs to λ, hence $v = v'/f$ belongs to λ/χ. □

DEFINITION 5.1.2 (*Asanuma modifications*). The upper line in (5.1) yields an affine modification $\mathcal{X}' \to \mathcal{X}$ called an *Asanuma modification of the first kind.* Its center is a reduced zero-dimensional subscheme of $\mathcal{X}$.

We call an *Asanuma modification of the second kind* an affine modification $\mathcal{X}'' \to \mathcal{X}$ of the cylinder $\mathcal{X} = X \times \mathbb{A}^1$ over a marked GDF surface $\pi: X \to B$ along divisor $\mathcal{D} = (f \circ \pi)^*(0)$ on $\mathcal{X}$ where $f \in \mathcal{O}_B(B) \setminus \{0\}$ with a one-dimensional center

$\mathbb{V}(I) \subset X \times \{0\}$ where $I = (f, v) \subset \mathcal{O}_{\mathcal{X}}(\mathcal{X})$. Due to the next lemma the latter modification results in a cylinder isomorphic to $\mathcal{X}$ over B.

5.1.3. Let $\pi: X \to B$ be a marked GDF μ_d-surface with a marking $z \in \mathcal{O}_B(B) \setminus \{0\}$. Recall (see Definition 4.1.3) that $\mathcal{X}(k)$ stands for the cylinder $\mathcal{X} = X \times \mathbb{A}^1$ equipped with a product μ_d-action where μ_d acts on the second factor via $(\zeta, v) \mapsto \zeta^k v$ for $v \in \mathbb{A}^1$ and $\zeta \in \mu_d$. By abuse of notation we still denote by π the μ_d-equivariant projection of the induced $\mathbb{A}^2$-fibration $\mathcal{X}(k) \to B$.

LEMMA 5.1.4. *In the notation of* 5.1.2–5.1.3 *consider an Asanuma modification of the second kind* $\mathcal{X}'' \to \mathcal{X}$. *Then the following hold.*

(a) *There is an isomorphism* $\mathcal{X}'' \cong_B \mathcal{X}$.

(b) *If* $\pi: X \to B$ *is a marked GDF* μ_d*-surface with a marking* $z \in \mathcal{O}_B(B) \setminus \{0\}$, $f = z$, *and* $\mathcal{X} = \mathcal{X}(k)$ *then* $\mathcal{X}'' = \mathcal{X}''(k-1)$.

(c) *Let things be in* (b). *Consider a second marked GDF* μ_d*-surface* $\pi': X' \to B$ *with the same marking* $z \in \mathcal{O}_B(B) \setminus \{0\}$, *and let* $\mathcal{X}' = X' \times \mathbb{A}^1$ *where* $\mathbb{A}^1 = \operatorname{Spec} \Bbbk[v']$. *Assume that for some natural* r *there is an equivariant isomorphism* $\varphi_r: \mathcal{X}(r) \xrightarrow{\cong_{\mu_d, B}} \mathcal{X}'(r)$ *such that* $\varphi_r^*(v') \equiv v \mod z^s$ *where* $s > d$. *Then for any* $k \in \mathbb{Z}$ *there is an equivariant isomorphism* $\varphi_k: \mathcal{X}(k) \xrightarrow{\cong_{\mu_d, B}} \mathcal{X}'(k)$ *such that* $\varphi_k^*(v') \equiv v \mod z^{s-d}$.

PROOF. (a) Indeed, the affine modification $\mathcal{X}'' \to \mathcal{X}$ amounts to

$$\mathcal{O}_{\mathcal{X}}(\mathcal{X}) \hookrightarrow \mathcal{O}_{\mathcal{X}''}(\mathcal{X}'') = \mathcal{O}_{\mathcal{X}}(\mathcal{X})[v/f] = \mathcal{O}_X(X)[v''] \text{ where } v'' = v/f. \tag{5.2}$$

(b) Under the assumptions of (b) one has $\zeta . v'' = \zeta^{k-1} v''$ for any $\zeta \in \mu_d$.

(c) Consider first the case $k = r - 1$. Let $I = (z, v) \subset \mathcal{O}_{\mathcal{X}}(\mathcal{X})$ and $I' = (z, v') \subset \mathcal{O}_{\mathcal{X}'}(\mathcal{X}')$. Under our assumptions one has $\varphi_r^*(I') = I$. By virtue of Lemma 1.4.3 the isomorphism φ_r lifts to an equivariant isomorphism, say, φ_{r-1} of the affine modifications of the cylinders $\mathcal{X}$ and $\mathcal{X}'$ along divisors $z^*(0)$ with the ideals I and I', respectively. By (b) this leads to a commutative diagram

$$\begin{array}{ccc} \mathcal{X}(r-1) & \xrightarrow[\cong_{\mu_d, B}]{\varphi_{r-1}} & \mathcal{X}'(r-1) \\ \downarrow & & \downarrow \\ \mathcal{X}(r) & \xrightarrow[\varphi_r]{\cong_{\mu_d, B}} & \mathcal{X}'(r) \end{array}$$

where the vertical arrows are the corresponding Asanuma modifications of the second kind and $\varphi_{r-1}^*(v'/z) \equiv v/z \mod z^{s-1}$. The sequence of characters $(\chi_k \in (\mu_d)^\vee)_{k \in \mathbb{Z}}$, $\chi_k(\zeta) = \zeta^k$, $\zeta \in \mu_d$, is periodic with period d. It follows that the sequences of cylinders $(\mathcal{X}(k))_{k \in \mathbb{Z}}$ and $(\mathcal{X}'(k))_{k \in \mathbb{Z}}$ are both periodic with period d. For $l \in \{0, \ldots, d-1\}$ the recursion on l yields the existence of isomorphisms $\varphi_{r-l}: \mathcal{X}(r-l) \xrightarrow{\cong_{\mu_d, B}} \mathcal{X}'(r-l)$ such that $\varphi_{r-l}^*(v'/z^l) \equiv v/z^l \mod z^{s-l}$. In particular, the congruence in (c) holds for these values of l. By periodicity, it holds for any $k \in \mathbb{Z}$. □

REMARK 5.1.5. Let $\varrho: X' \to X$ be a fibered modification as in Lemma 5.1.1. Consider the product modification of cylinders $\sigma = \varrho \times \mathrm{id}: \mathcal{X}' \to \mathcal{X}$ followed by the Asanuma modification of the second kind $\mathcal{X}'' \to \mathcal{X}'$ with $f = z$. This yields an affine modification $\mathcal{X}'' \to \mathcal{X}$ factorized as in Remark 1.4.2.2. Identifying $\mathcal{X}'$ and $\mathcal{X}''$ via an isomorphism as in Lemma 5.1.4 gives an Asanuma modification of the first kind

$\tilde{\varrho}: \mathcal{X}' \to \mathcal{X}$ such that $\varrho = \tilde{\varrho}|_{X' \times \{0\}}$. Under this correspondence the conclusions of Lemmas 5.1.1(b) and 5.1.4(c) agree in the μ_d-equivariant setting.

5.2. Rigidity of cylinders under deformations of GDF surfaces

Form Lemma 5.1.1 we deduce the following corollary.

COROLLARY 5.2.1. (a) *Consider a marked GDF μ_d-surface $\pi: X \to B$ along with a trivializing μ_d-equivariant sequence (2.7) of fibered modifications, see Corollary 2.4.6(b). Given $l \in \{1, \ldots, m\}$ and $k \in \mathbb{Z}$ the fibered modification $\varrho_l: X_l \to X_{l-1}$ as in (2.7) along divisor $z^*(0)$ with a center, say, I_l induces a μ_d-equivariant Asanuma modification of the first kind $\tilde{\varrho}_l: \mathcal{X}_l(k) \to \mathcal{X}_{l-1}(k+1)$ over B along divisor $z^*(0)$ on $\mathcal{X}_{l-1}$ with center $J_l = (I_l, v)$, cf. Lemma 5.1.1.*

(b) *Consequently, (2.7) yields a sequence of μ_d-equivariant affine modifications*

$$\mathcal{X}_m(-m) \xrightarrow{\tilde{\varrho}_m} \mathcal{X}_{m-1}(-m+1) \longrightarrow \ldots \longrightarrow \mathcal{X}_1(-1) \xrightarrow{\tilde{\varrho}_1} \mathcal{X}_0(0) = (B \times \mathbb{A}^2)(0)\,. \tag{5.3}$$

PROOF. The statement of (a) follows by Lemma 5.1.1 and (b) is immediate from (a). □

The next theorem is the main result of this section.

THEOREM 5.2.2. *Let $\pi: X \to B$ and $\pi': X' \to B$ be two marked GDF μ_d-surfaces with the same μ_d-quasi-invariant marking $z \in \mathcal{O}_B(B) \setminus \{0\}$ of weight 1. Assume that for some trivializing μ_d-equivariant completions $(\hat{X}, \hat{D})$ and $(\hat{X}', \hat{D}')$ of X and X', respectively, the graph divisors $\mathcal{D}(\hat{\pi})$ and $\mathcal{D}(\hat{\pi}')$ are μ_d-equivariantly isomorphic (see Definition 2.3.11). Then for any $k \in \mathbb{Z}$ there is a μ_d-equivariant isomorphism $\mathcal{X}(k) \cong_{\mu_d, B} \mathcal{X}'(k)$. In particular, $\mathcal{X}(0) \cong_{\mu_d, B} \mathcal{X}'(0)$.*

PROOF. The trivializing sequences (5.3) associated with the GDF surfaces X and X', respectively, start both with the same product $\mathcal{X}_0(0) = (B \times \mathbb{A}^2)(0) = \mathcal{X}'_0(0)$. Using Proposition 5.2.3 below one shows by induction on l that for any $l = 0, \ldots, m$ there is a μ_d-equivariant isomorphism $\varphi_l: \mathcal{X}_l(-l) \xrightarrow{\cong_{\mu_d, B}} \mathcal{X}'_l(-l)$. In particular, for $l = m$ one obtains an isomorphism $\varphi_m: \mathcal{X}(-m) \xrightarrow{\cong_{\mu_d, B}} \mathcal{X}'(-m)$. Then by Lemma 5.1.4(c) for any $k \in \mathbb{Z}$ one gets a μ_d-equivariant isomorphism $\mathcal{X}(k) \cong_{\mu_d, B} \mathcal{X}'(k)$. □

The following proposition provides the inductive step in the proof of Theorem 5.2.2.

PROPOSITION 5.2.3. *Under the assumptions of Theorem 5.2.2 suppose that for some $l \in \{0, \ldots, m-1\}$ there exists a μ_d-equivariant isomorphism $\psi_l: \mathcal{X}_l(-l) \xrightarrow{\cong_{\mu_d, B}} \mathcal{X}'_l(-l)$ such that*

(i_l) *the induced correspondence between the special fiber components of π_l and π'_l is the restriction of the isomorphism $\mathcal{D}(\hat{\pi}) \xrightarrow{\cong} \mathcal{D}(\hat{\pi}')$;*

(ii_l) *$\psi_l^*(v'_l) \equiv v_l \bmod z^s$ where $s > 0$ and v_l (v'_l, respectively) is an affine coordinate in the $\mathbb{A}^1$-factor of the cylinder $\mathcal{X}_l(-l)$ ($\mathcal{X}'_l(-l)$, respectively).*

Then there exists a μ_d-equivariant isomorphism $\psi_{l+1}: \mathcal{X}_{l+1}(-l-1) \xrightarrow{\cong_{\mu_d, B}} \mathcal{X}'_{l+1}(-l-1)$ such that

(i_{l+1}) *the induced correspondence between the special fiber components of π_{l+1} and π'_{l+1} is the restriction of the isomorphism $\mathcal{D}(\hat{\pi}) \xrightarrow{\cong} \mathcal{D}(\hat{\pi}')$;*

(ii$_{l+1}$) $\psi_{l+1}^*(v'_{l+1}) \equiv v_{l+1} \bmod (z^{s-1})$.

PROOF. The morphism $\varrho_{l+1}: X_{l+1} \to X_l$ in (2.7) is a μ_d-equivariant affine modification along the reduced principal divisor $\mathbb{V}(z) = z^*(0)$ on X_l with a reduced center I where $\mathbb{V}(I)$ is the union of a finite set $\Sigma \subset X_l$ and the components of $\mathbb{V}(z)$ disjoint from Σ, cf. Remark 2.4.2. Notice that Σ is contained in the union of the top level components F of $\mathbb{V}(z)$. Let $\mathfrak{F}_\Sigma$ be the set of the top-level components $\mathcal{F} = F \times \mathbb{A}^1$ which meet $\Sigma \times \{0\}$. By Lemma 5.1.1, ϱ_{l+1} induces a μ_d-equivariant Asanuma modification of the first kind $\tilde{\varrho}_{l+1}: \mathcal{X}_{l+1}(-l-1) \to \mathcal{X}_l(-l)$ with the principal divisor $\mathbb{V}(z) \times \mathbb{A}^1$ and center $\mathbb{V}(I) \times \{0\} \subset \mathcal{X}_l(-l)$ consisting of a μ_d-invariant finite set $\Sigma \times \{0\}$ and a μ_d-invariant union C of curves isomorphic to $\mathbb{A}^1$ such that $C_\mathcal{F} = C \cap \mathcal{F}$ is given by equation $v_l = 0$ in each component $\mathcal{F} = F \times \mathbb{A}^1 \notin \mathfrak{F}_\Sigma$. Thus, $C \subset \{z = v_l = 0\}$ in $\mathcal{X}_l(-l)$. For any $\mathcal{F} \in \mathfrak{F}_\Sigma$ we let

$$\Sigma_\mathcal{F} = \mathcal{F} \cap (\Sigma \times \{0\}) = \{x_1, \ldots, x_{M(\mathcal{F})}\}. \tag{5.4}$$

There is a similar collection of objects related with X' instead of X. In particular, one has a modification $\tilde{\varrho}'_{l+1} : \mathcal{X}'_{l+1}(-l-1) \to \mathcal{X}'_l(-l)$ with the divisor $\mathbb{V}(z) \times \mathbb{A}^1$ and the center $\mathbb{V}(I') \times \{0\}$ consisting of a μ_d-invariant finite set $\Sigma' \times \{0\}$ and a μ_d-invariant union C' of curves $C'_{\mathcal{F}'} \cong \mathbb{A}^1$.

By virtue of (i$_l$) the μ_d-equivariant isomorphism $\mathcal{D}(\pi) \xrightarrow{\cong} \mathcal{D}(\pi')$ of graph divisors yields a one-to-one correspondence $\mathcal{F} \rightsquigarrow \mathcal{F}'$ between the components in $\mathfrak{F}_\Sigma$ and in $\mathfrak{F}'_{\Sigma'}$ so that (see (5.4))

$$M(\mathcal{F}) = \operatorname{card} \Sigma_\mathcal{F} = \operatorname{card} \Sigma_{\mathcal{F}'} = M(\mathcal{F}') \qquad \forall \mathcal{F} \in \mathfrak{F}_\Sigma .$$

By virtue of (ii$_l$), ψ_l sends the pair $(\mathcal{X}_l(-l), \mathbb{V}(z) \times \mathbb{A}^1)$ to the pair $(\mathcal{X}'_l(-l), \mathbb{V}(z) \times \mathbb{A}^1)$ and C to C', but not in general $\Sigma \times \{0\}$ to $\Sigma' \times \{0\}$. To get a bijection between the centers Σ and Σ' of modifications we will replace ψ_l by a composition $\varphi_l \circ \psi_l$ with a suitable μ_d-equivariant automorphism $\varphi_l \in \operatorname{SAut}_B(\mathcal{X}'_l(-l))$.

Let $(z, u, v) = (z, u_l, v_l)$ be μ_d-quasi-invariant natural coordinates in the standard affine chart $U_F \times \mathbb{A}^1$ about $\mathcal{F}$ where $\mathcal{F} \in \mathfrak{F}_\Sigma$, see Definition 3.3.1. For a point x_ν in (5.4) one has $x_\nu = (0, u(x_\nu), 0)$. Similarly, for $\mathcal{F}' = F' \times \mathbb{A}^1 = \psi_l(\mathcal{F})$ consider the standard affine chart $U_{F'} \times \mathbb{A}^1$ about $\mathcal{F}'$ with natural coordinates $(z, u', v') = (z, u'_l, v'_l)$. Let

$$\mathcal{F}' \cap (\mathbb{V}(I') \times \{0\}) = \mathcal{F}' \cap (\Sigma' \times \{0\}) = \{x'_1, \ldots, x'_{M(\mathcal{F})}\}$$

where $x'_\nu = (0, u'(x'_\nu), 0)$.

Let μ_e with $e = e(\mathcal{F}) > 1$ be the stabilizer of $\mathcal{F} \in \mathfrak{F}_\Sigma$ in μ_d. Then $M(\mathcal{F}) \equiv 1 \mod e$ if $\bar{0} \in \Sigma_\mathcal{F}$ and $M(\mathcal{F}) \equiv 0 \mod e$ otherwise. Since $\psi_l: \mathcal{F} \to \mathcal{F}'$ is μ_d-equivariant one has $e(\mathcal{F}') = e(\mathcal{F})$. Since also $M(\mathcal{F}') = M(\mathcal{F})$ it follows that $\bar{0} \in \Sigma_\mathcal{F}$ if and only if $\bar{0} \in \Sigma_{\mathcal{F}'}$.

By (ii$_l$) one obtains

$$\psi_l(x_\nu) =: x''_\nu = (0, u'(x''_\nu), 0) \in \mathcal{F}', \quad \nu = 1, \ldots, M(\mathcal{F}).$$

Suppose that $e(\mathcal{F}) > 1$ and $x_\nu = \bar{0} \in \Sigma_\mathcal{F}$. Since $\psi_l: \mathcal{F} \to \mathcal{F}'$ is μ_d-equivariant is sends the orbits to the orbits. It follows that $x''_\nu = \psi_l(x_\nu) = \bar{0} \in \Sigma_{\mathcal{F}'}$. Up to renumbering one may assume in this case that $x'_\nu = x''_\nu = \bar{0}$.

CLAIM 1. There exists a μ_d-equivariant automorphism $\varphi_l \in \operatorname{SAut}_{\mu_d, B}(\mathcal{X}'_l(-l))$ as in Definition 4.1.3 with prescribed μ_d-equivariant s-jets in the points x''_ν chosen so that

(j) $\varphi_l(\mathcal{F}') = \mathcal{F}'$ for every component $\mathcal{F}' = F' \times \mathbb{A}^1$ of the divisor $z^*(0)$ on $\mathcal{X}'_l(-l)$;

(jj) $\varphi_l^*(v') \equiv v' \mod z^s$ near $\mathcal{F}'$ $\forall \mathcal{F}' \notin \mathfrak{F}_{\Sigma'}$;

(jjj) up to reordering, $\varphi_l(x''_\nu) = x'_\nu$, $\nu = 1, \ldots, M(\mathcal{F}')$ $\forall \mathcal{F}' \in \mathfrak{F}_{\Sigma'}$;

(jv) the s-jets of $\varphi_l^*(v')$ and v' at x''_ν coincide for any $\nu = 1, \ldots, M(\mathcal{F}')$, $\forall \mathcal{F}' \in \mathfrak{F}_{\Sigma'}$.

PROOF OF CLAIM 1. Condition (j) holds for any $\varphi \in \mathrm{SAut}_B(\mathcal{X}'_l)$, cf. Proposition 3.2.2. Due to Theorem 4.1.4 the surface X' verifies the μ_d-equivariant condition $\mathrm{RF}(l, -l, s)$. Therefore, one can choose $\varphi_l \in \mathrm{SAut}_{\mu_d, B}(\mathcal{X}'_l(-l))$ verifying conditions (α_1), (α_2), and (β) of Definitions 4.1.2 and 4.1.3 with a suitable data. This yields (jj), and as well (jjj) and (jv) in the case where either $e(\mathcal{F}') = 1$ or $\bar{0} \notin \Sigma_{\mathcal{F}'}$.

In the remaining case one has $e(\mathcal{F}') > 1$ and $\bar{0} = x'_\nu \in \Sigma_{\mathcal{F}'}$. By the observations preceding the claim one may assume that $x'_\nu = x''_\nu = \bar{0}$ and the s-jet of φ_l at $\bar{0}$ is the s-jet of the identity. This yields (jjj) and (jv) also in the remaining case. Now the claim follows.

Due to (j)–(jjj) the composition $\tilde{\psi}_l := \varphi_l \circ \psi_l$ sends the center and the divisor of $\tilde{\varrho}_{l+1}$ to the center and the divisor of $\tilde{\varrho}'_{l+1}$. By Lemma 1.4.3, $\tilde{\psi}_l$ lifts to a μ_d-equivariant isomorphism $\psi_{l+1}: \mathcal{X}_{l+1}(-l-1) \xrightarrow{\cong_{\mu_d, B}} \mathcal{X}'_{l+1}(-l-1)$. The proof ends due to the following

CLAIM 2. ψ_{l+1} satisfies conditions (i_{l+1}) and (ii_{l+1}).

PROOF OF CLAIM 2. Due to conditions (i_l) for ψ_l and (j) for φ_l the isomorphism $\mathcal{D}(\hat{\pi})_{\le l} \xrightarrow{\cong} \mathcal{D}(\hat{\pi}')_{\le l}$ induced by ψ_{l+1} coincides with the restriction of the given isomorphism $\mathcal{D}(\hat{\pi}) \xrightarrow{\cong} \mathcal{D}(\hat{\pi}')$. The same holds for the induced isomorphism $\mathcal{D}(\hat{\pi})_{\le l+1} \xrightarrow{\cong} \mathcal{D}(\hat{\pi}')_{\le l+1}$ after a suitable renumbering of the points $x'_1, \ldots, x'_{M(\mathcal{F}')}$ on each component $\mathcal{F}' \in \mathfrak{F}'$. This gives (i_{l+1}).

For any special fiber component $\mathcal{F}$ in $\mathcal{X}_{l+1}(-l-1)$ of level $\le l$ condition (ii_{l+1}) holds due to (ii_l), (jj), and the equalities $v_{l+1} = v_l/z$, $v'_{l+1} = v'_l/z$. It holds as well for $\mathcal{F}$ of the top level $l+1$ due to (ii_l), (jv), and the same equalities. □

5.3. Rigidity of cylinders under deformations of $\mathbb{A}^1$-fibered surfaces

Using Theorem 5.2.2 we obtain our second main result.

THEOREM 5.3.1. *Let $\pi: Y \to C$ and $\pi': Y' \to C$ be two $\mathbb{A}^1$-fibered normal affine surfaces over a smooth affine curve C. Let $\hat{Y} \to \hat{C}$ be an SNC completion of the minimal resolution of singularities of Y, and let $\hat{D}_{\mathrm{ext}}$ be the extended divisor of this completion. Let a pair $(\hat{Y}', \hat{D}'_{\mathrm{ext}})$ plays the same role for Y'. Suppose that*

- *the degenerate fibers of π and π' are situated over the same points $p_1, \ldots, p_t \in C$;*
- *for $i = 1, \ldots, t$ the corresponding rooted fiber trees $\Gamma_{p_i}(\pi)$ and $\Gamma_{p_i}(\pi')$ are isomorphic*[1]*;*
- *making similar contractions in $\hat{D}_{\mathrm{ext}}$ and $\hat{D}'_{\mathrm{ext}}$ one can reduce both $\hat{Y}$ and $\hat{Y}'$ to the product $\hat{C} \times \mathbb{P}^1$ with the same distinguished "section at infinity" $\hat{C} \times \{\infty\}$.*

[1]This condition ensures that the corresponding fiber components of π and π' have the same multiplicities. Indeed, under our assumptions the isomorphism respects the feathers along with the associated bridges.

Then the cylinders $Y \times \mathbb{A}^1$ and $Y' \times \mathbb{A}^1$ are isomorphic over C.

PROOF. Applying a suitable cyclic Galois base change $B \to C$ of order d ramified over the points $p_1, \ldots, p_t \in C$ one can replace the $\mathbb{A}^1$-fibered surfaces $\pi: Y \to C$ and $\pi': Y' \to C$ by two marked GDF μ_d-surfaces $X \to B$ and $X' \to B$, respectively, with the same μ_d-quasi-invariant marking $z \in \mathcal{O}_B(B) \setminus \{0\}$, see Lemma 2.1.3 and Remark 2.1.4. Due to our assumptions the extended graphs and the fiber trees of the special fibers of suitable μ_d-equivariant pseudominimal completions $\bar{X}$ and $\bar{X}'$ of the GDF surfaces X and X' are isomorphic under a μ_d-equivariant isomorphism.[2] Moreover, these surfaces admit trivializing μ_d-equivariant completions $\hat{X}$ and $\hat{X}'$, respectively, verifying the assumptions of Theorem 5.2.2. Due to this theorem there is a μ_d-equivariant isomorphism $\mathcal{X}(0) \cong_{\mu_d, B} \mathcal{X}'(0)$. Passing to the quotients $\mathcal{X}(0)/\mu_d = Y \times \mathbb{A}^1$ and $\mathcal{X}'(0)/\mu_d = Y' \times \mathbb{A}^1$ yields a desired C-isomorphism $Y \times \mathbb{A}^1 \cong_C Y' \times \mathbb{A}^1$. □

COROLLARY 5.3.2. *Let C be a smooth affine curve with marked points $p_1, \ldots, p_t \in C$. Consider the collection $\mathfrak{H} = \mathfrak{H}(C, p_1, \ldots, p_t)$ of all the $\mathbb{A}^1$-fibered normal affine surfaces $\pi: X \to C$ such that π restricted over $C \setminus \{p_1, \ldots, p_t\}$ is the projection of a trivial line bundle. Then the set of C-isomorphism classes of cylinders $\mathcal{X} = X \times \mathbb{A}^1$, where X runs over $\mathfrak{H}$, is at most countable.*

PROOF. Indeed, the set of the isomorphism classes of finite trees is countable. The same is true for the set of all ordered t-tuples of such trees as in Theorem 5.3.1. Now the assertion follows from this theorem. □

REMARK 5.3.3. Let $\pi: X \to C$ and $\pi': X' \to C'$ be $\mathbb{A}^1$-fibered normal affine surfaces. If $C \ncong \mathbb{A}^1$ then any isomorphism of cylinders $\varphi: \mathcal{X} \xrightarrow{\cong} \mathcal{X}'$ fits in a commutative diagram

$$\begin{array}{ccc} \mathcal{X} & \xrightarrow[\cong]{\varphi} & \mathcal{X}' \\ \downarrow & & \downarrow \\ C & \xrightarrow[\psi]{\cong} & C' \end{array}$$

where ψ is an isomorphism (cf. also Lemma 6.3.4). For instance, the isomorphism type of the cylinder $\mathcal{X}$ over the surface $X = (\mathbb{A}^1 \setminus \{k \text{ points}\}) \times \mathbb{A}^1$ depends essentially on the isomorphism type of the factor $\mathbb{A}^1 \setminus \{k \text{ points}\}$ (see [**36**, 9.10.1]).

5.4. Rigidity of line bundles over affine surfaces

In unpublished notes [**11**] kindly provided to us by the authors the study of cylinders over affine surfaces is extended to the total spaces of line bundles over affine surfaces. Theorem 5.4.8 below is an analog of Theorem 5.2.2 in this wider context. We do not use this extension in the sequel, so, we just indicate the necessary modifications in the proof of Theorem 5.2.2.

NOTATION 5.4.1. Let X be an affine algebraic variety. For a Cartier divisor $T \in \operatorname{CDiv} X$ we let $\pi^T: \mathcal{X}^T \to X$ be the associated line bundle on X with a zero section $Z^T \subset \mathcal{X}^T$.

[2]Such completions are not unique. However, the deformation parameters are irrelevant for the combinatorial invariants such as the extended graph and the fiber trees.

DEFINITION 5.4.2. Let $D \in \operatorname{CDiv} X$ be a reduced effective Cartier divisor on X. By an *Asanuma modification of the second kind* of $\mathcal{X}^T$ we mean an affine modification $\sigma^D: \mathcal{X}^{T,D} \to \mathcal{X}^T$ along the principal divisor $\mathcal{D}^T = (\pi^T)^*(D)$ on $\mathcal{X}^T$ with center $\mathcal{D}^T \cdot Z^T$.

We have the following analogue of Lemma 5.1.1 (corresponding to the case $T \sim 0$).

LEMMA 5.4.3. *In Notation 5.4.1, $\pi^{T,D} = \pi^T \circ \sigma^D: \mathcal{X}^{T,D} \to X$ admits a structure of a line bundle such that $\mathcal{X}^{T,D} \cong_X \mathcal{X}^{T-D}$.*

PROOF. Choose an open covering $X = \bigcup_i U_i$ such that

- $D \cap U_i = f_i^*(0)$ and $T \cap U_i = \operatorname{div} h_i$ where $f_i \in \mathcal{O}_{U_i}(U_i)$ and $h_i \in \operatorname{Frac} \mathcal{O}_{U_i}(U_i)$.

Then

- $\alpha_{i,j} = f_j/f_i$, $\beta_{i,j} = h_j/h_i \in \mathcal{O}^\times_{U_{i,j}}(U_{i,j})$ where $U_{i,j} = U_i \cap U_j$, are Čech 1-cocycles on X associated with the line bundles $\mathcal{X}^D \to X$ and $\mathcal{X}^T \to X$, respectively.

Letting $V_i = (\pi^T)^{-1}(U_i)$ there are local trivializations $V_i \cong_{U_i} U_i \times \mathbb{A}^1$ of $\pi^T: \mathcal{X}^T \to X$ where $\mathbb{A}^1 = \operatorname{Spec} \Bbbk[v_i]$ with $v_j = \beta_{i,j} v_i$ over $U_{i,j}$. Consider the restriction $V_i' \to V_i$ of the morphism $\sigma: \mathcal{X}^{T,D} \to X^T$ over V_i induced by the natural inclusion

$$\mathcal{O}_{V_i}(V_i) \hookrightarrow \mathcal{O}_{V_i'}(V_i') = \mathcal{O}_{V_i}(V_i)[v_i/f_i].$$

One has $V_i' \cong_{U_i} U_i' \times \mathbb{A}^1$ where $\mathbb{A}^1 = \operatorname{Spec} \Bbbk[v_i']$ with $v_i' = v_i/f_i$. This defines local trivializations of the projection $\pi^{T,D}: \mathcal{X}^{T,D} \to X$, hence a structure of a line bundle on $\mathcal{X}^{T,D}$ over X. Note that $v_j' = (\alpha_{i,j}^{-1}\beta_{i,j}) v_i'$ where $\{\alpha_{i,j}^{-1}\beta_{i,j}\}$ is a Čech 1-cocycle on X associated with the line bundle $\pi^{T-D}: \mathcal{X}^{T-D} \to X$. □

NOTATION 5.4.4. Let X be an affine variety acted upon by a finite group G, and let $T, D \in \operatorname{Div}(X)$ be G-invariant divisors where D is reduced. Then the line bundle $\mathcal{X}^T \to X$ admits a G-linearization, that is, a structure of a G-equivariant line bundle. This structure is not unique, in general. It is defined modulo the multiplication by a character, see, e.g., [**59**]. Choosing a G-linearization, say, $\mathcal{X}^T(1) \to X$ with the corresponding linear equivariant G-action $\varphi: (g, v) \mapsto g.v$ on $\mathcal{X}^T$, for a character $\chi \in G^\vee$ consider a new such action $\varphi^\chi: (g, v) \mapsto \chi(g) \cdot g.v$. This yields a new G-linearization denoted by $\mathcal{X}^T(\chi) \to X$.

In the case of a cyclic group $G = \mu_d$, fixing a primitive character χ of μ_d we write $\mathcal{X}^T(k) \to X$ for the μ_d-linearization on $\mathcal{X}^T \to X$ associated with the character χ^k. With this notation, $\mathcal{X}^T(0) \to X$ corresponds to the given G-linearization. Clearly, the sequence $(\mathcal{X}^T(k))_{k \in \mathbb{Z}}$ is periodic with period d. For any G-invariant divisors $T_1, T_2 \in \operatorname{Div}(B)$ and any characters $\chi, \lambda \in G^\vee$ there is a G-equivariant isomorphism $\mathcal{X}^{T_1}(\chi) \otimes \mathcal{X}^{T_2}(\lambda) \cong_X \mathcal{X}^{T_1+T_2}(\chi\lambda)$.

In the sequel we need the following simple lemma.

LEMMA 5.4.5. *Let B be a smooth affine curve acted upon by a finite group G, and let $\xi: L \to B$ be a line bundle over B which admits a G-linearization. Then for any $b_1, \dots, b_n \in B$ there are a G-invariant open set U containing these points and a G-equivariant trivialization of $\xi|_U$.*

PROOF. It suffices to find a nonzero G-stable (that is, G-equivariant) rational section $s: B \to L$ of ξ which has neither pole nor zero in $b_1, \dots, b_n$ and to set

$U = B \setminus \operatorname{supp}(\operatorname{div} s)$. Given any nonzero G-stable rational section $s_0: B \to L$ of ξ one can find a G-invariant rational function $f \neq 0$ on B such that $\operatorname{div} f$ restricts to $\operatorname{div} s$ on $b_1, \ldots, b_n$. Then $s = s_0/f$ is a desired G-stable section of ξ. □

NOTATION 5.4.6. Let $\pi: X \to B$ be a marked GDF μ_d-surface over a smooth affine curve B with a μ_d-quasi-invariant marking $z \in \mathcal{O}_B(B) \setminus \{0\}$ of weight 1. Then the principal divisor $D = z^*(0) \in \operatorname{Div}(B)$ is μ_d-invariant. Given a μ_d-invariant divisor $T \in \operatorname{Div}(B)$ consider the line bundle $\mathcal{X}^{T^*} \to X$ where $T^* = \pi^*(T) \in \operatorname{Div}(X)$. By abuse of notation we let $\mathcal{X}^T = \mathcal{X}^{T^*}$. If $\xi: L \to B$ is the line bundle associated with T then $\mathcal{X}^T \to X$ is induced by ξ via the morphism $\pi: X \to B$. Hence both ξ and $\mathcal{X}^T \to X$ admit μ_d-linearizations such that the natural morphism $\mathcal{X}^T \to L$ is μ_d-equivariant. Choosing such a μ_d-linearization of ξ and the one of $\mathcal{X}^T \to X$ we observe that $L(k)$ naturally corresponds to $\mathcal{X}^T(k)$.

There is the following equivariant version of Lemma 5.4.3.

LEMMA 5.4.7. *Let things be as in 5.4.6. Then for any $k \in \mathbb{Z}$ there exists a μ_d-action on $\mathcal{X}^{T,D}$ and a μ_d-equivariant isomorphism of line bundles $\mathcal{X}^{T,D} \cong_{\mu_d,B} \mathcal{X}^{T-D}(k-1)$ such that the induced morphism $\sigma^D: \mathcal{X}^{T-D}(k-1) \to \mathcal{X}^T(k)$ is μ_d-equivariant.*

PROOF. The μ_d-action on $\mathcal{X}^T(k)$ stabilizes the divisor $\mathcal{D}^T = (\pi^T)^*(D) \in \operatorname{Div}(\mathcal{X}^T)$ and the center $\mathcal{D}^T \cdot Z^T$ of the affine modification $\sigma^D: \mathcal{X}^{T,D} \to \mathcal{X}^T(k)$. By [**47**, Cor. 2.2] (see Lemma 1.4.3) it lifts to a μ_d-action on $\mathcal{X}^{T,D}$ making σ^D equivariant.

Choose a trivializing open set $U \subset B$ as in Lemma 5.4.5, and let $V = \pi_X^{-1}(U) \subset X$. Then $\mathcal{X}^T(k) \to X$ admits over V a μ_d-equivariant trivialization $\mathcal{X}^T(k)|_V \cong_{\mu_d,V} (V \times \mathbb{A}^1)(k+m)$ where $\mathbb{A}^1 = \operatorname{Spec} \Bbbk[v]$ and m is the weight of an equivariant trivialization of $\mathcal{X}^T(0)|_V$. Recall that $D = \operatorname{div} z$ where z has weight 1 and there is a natural isomorphism $\mathcal{X}^{T,D}|_V \cong_V V \times \mathbb{A}^1$ where $\mathbb{A}^1 = \operatorname{Spec} \Bbbk[v/z]$ compatible with an isomorphism $\mathcal{X}^{T,D} \cong_B \mathcal{X}^{T-D}$ of Lemma 5.4.3 (see the proof of this lemma). This gives an equivariant trivialization $\mathcal{X}^{T-D}|_V \cong_{\mu_d,V} (V \times \mathbb{A}^1)(k+m-1)$ and shows that the induced μ_d-action on $\mathcal{X}^{T-D}$ has weight $k-1$. Now the assertions follow. □

The following result is an analog of Theorem 5.2.2 in our more general setting.

THEOREM 5.4.8. *Let $\pi_X: X \to B$ and $\pi_Y: Y \to B$ be two marked GDF μ_d-surfaces over B with the same μ_d-quasi-invariant marking $z \in \mathcal{O}_B(B) \setminus \{0\}$ of weight 1. Assume that for some trivializing μ_d-equivariant completions $(\hat{X}, \hat{D}_X)$ and $(\hat{Y}, \hat{D}_Y)$ the graph divisors $\mathcal{D}(\hat{\pi}_X)$ and $\mathcal{D}(\hat{\pi}_Y)$ are μ_d-equivariantly isomorphic. Let $T \in \operatorname{Div}(B)$ be a μ_d-invariant divisor. Then for any $k \in \mathbb{Z}$ there is a μ_d-equivariant isomorphism $\mathcal{X}^T(k) \cong_{\mu_d,B} \mathcal{Y}^T(k)$.*

In the proof we use an analog of the Asanuma modification of the first kind for line bundles over surfaces (see Definition 5.4.10 below). Let us introduce the following notation.

NOTATION 5.4.9. Let $z^{-1}(0) = \{b_1, \ldots, b_n\} \subset B$. Consider a trivializing sequence (2.7) of fibered modifications $\varrho_{l+1}: X_{l+1} \to X_l$, $l = 0, \ldots, m$. Let T be a μ_d-invariant divisor on B, and let f be a rational μ_d-quasi-invariant function on B such that $(\operatorname{div} f)(b_i) = -T(b_i)$, $i = 1, \ldots, n$. Then T and $T + \operatorname{div} f$ represent the same class in $\operatorname{Pic} B$. Replacing T by $T + \operatorname{div} f$ we may assume that $b_i \notin \operatorname{supp} T$ $\forall i = 1, \ldots, n$.

For every $l = 0, \dots, m$ we let $T_l = \pi_l^*(T) \in \mathrm{Div}(X_l)$. Since $T_{l+1} = \varrho_{l+1}^*(T_l)$ the modification $\varrho_{l+1}: X_{l+1} \to X_l$ induces an affine modification $\varrho_{l+1}^T: \mathcal{X}_{l+1}^{T_{l+1}} \to \mathcal{X}_l^{T_l}$ which fits in the commutative diagram

(5.5)

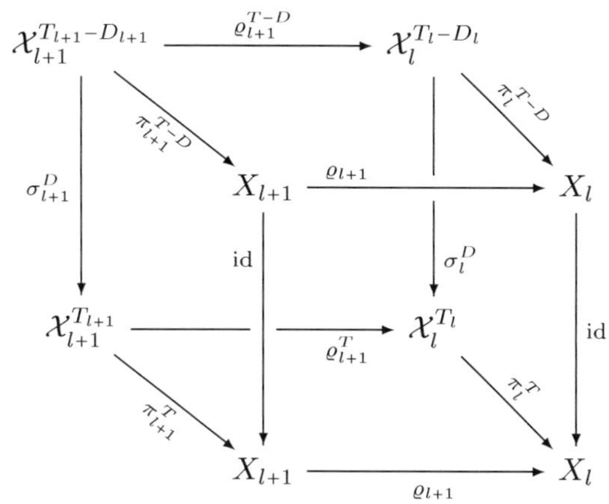

For a fiber component $F_i \subset D_l$ we let C_i be the intersection of F_i with center of the modification $\varrho_{l+1}: X_{l+1} \to X_l$. Then $\varrho_{l+1}^T: \mathcal{X}_{l+1}^{T_{l+1}} \to \mathcal{X}_l^{T_l}$ is an affine modification along divisor $\mathcal{D}_l = \bigcup_i \mathcal{F}_i$ with center $\mathcal{C}_l = \bigcup_i \mathcal{C}_i$ where $\mathcal{F}_i = (\pi^{T_l})^{-1}(F_i) \cong F_i \times \mathbb{A}^1 \cong \mathbb{A}^2$ and $\mathcal{C}_i \cong C_i \times \mathbb{A}^1 \subset F_i \times \mathbb{A}^1$. There is an alternative: either

(i) C_i is finite, or
(ii) $C_i = F_i$.

In case (i), F_i is a top level component. In case (ii), $X_{l+1} \to X_l$ ($\mathcal{X}_{l+1}^{T_{l+1}} \to \mathcal{X}_l^{T_l}$, respectively) is an isomorphism near F_i (near $\mathcal{F}_i$, respectively).

DEFINITION 5.4.10. By analogy, we call an *Asanuma modification of the first kind* the birational morphism

$$\kappa_{l+1}: \mathcal{X}_{l+1}^{T_{l+1}-D_{l+1}} \to \mathcal{X}_l^{T_l}$$

where $\kappa_{l+1} = \varrho_{l+1}^T \circ \sigma_{l+1}^D = \sigma_l^D \circ \varrho_{l+1}^{T-D}$ is the diagonal composition of morphisms in the back square of (5.5). Then κ_{l+1} is an affine modification along divisor $\mathcal{D}_l = \bigcup_i \mathcal{F}_i$ on $\mathcal{X}_l^{T_l}$ with center $\bigcup_i (C_i \times \{0\})$. In case (i), $C_i \times \{0\} \subset \mathcal{C}_i \cong \mathbb{A}^2$ is zero-dimensional, while in case (ii) this is just the coordinate axis $v = 0$ in $\mathcal{C}_i \cong \mathbb{A}^2$. Due to Lemma 5.4.7 and by analogy with (5.3) one has the following sequence of equivariant Asanuma modifications of the first kind:

$$\mathcal{X}_m^{T_m - mD_m}(-m) \xrightarrow{\tilde{\varrho}_m} \dots \longrightarrow \mathcal{X}_2^{T_2-2D_2}(-2) \longrightarrow \mathcal{X}_1^{T_1-D_1}(-1) \xrightarrow{\tilde{\varrho}_1} \mathcal{X}_0^{T_0}(0)\,. \tag{5.6}$$

PROOF OF THEOREM 5.4.8. For the given GDF surfaces $\pi_X: X \to B$ and $\pi_Y: Y \to B$, consider the corresponding sequences (5.6) starting with the same line bundle $\mathcal{X}_0^{T_0}(0) = (B \times \mathbb{A}^1)^{T_0}(0) = \mathcal{Y}_0^{T_0}(0)$. One may suppose that μ_d acts trivially on the factor $\mathbb{A}^1$. Using Proposition 5.4.11 below with $s > m$ it follows by induction that for $l = 0, \dots, m$ there is a (non-linear, in general) μ_d-equivariant isomorphism

$$\mathcal{X}_l^{T_l - lD_l}(-l) \cong_{\mu_d, B} \mathcal{Y}_l^{T_l - lD_l}(-l)$$

which sends the zero section $Z(\mathcal{X}_l^{T_l - lD_l}(-l))$ of the first line bundle to such a section of the second one. Replacing T by $T + mD$ one obtains for $l = m$,

$$\mathcal{X}^T(-m) = \mathcal{X}_m^{T_m}(-m) \xrightarrow{\varphi} \mathcal{Y}_m^{T_m}(-m) = \mathcal{Y}^T(-m)\,,$$

where φ is a (μ_d, B)-isomorphism respecting the zero sections $Z(\mathcal{X}^T(-m))$ and $Z(\mathcal{Y}^T(-m))$ and the divisors $\mathcal{D}^T(\mathcal{X}^T)$ and $\mathcal{D}^T(\mathcal{Y}^T)$. Hence φ respects also the centers $\mathcal{D}^T(\mathcal{X}^T)\cdot Z(\mathcal{X}^T(-m))$ and $\mathcal{D}^T(\mathcal{Y}^T)\cdot Z(\mathcal{Y}^T(-m))$ of the Asanuma modifications of the second kind. Applying these modifications on both sides, by Lemma 5.4.7 we decrease by 1 the weights of the μ_d-actions. Due to Lemma 1.4.3, φ admits a lift to a (μ_d, B)-isomorphism $\tilde{\varphi}$ fitting in the commutative diagram

$$\begin{array}{ccc} \mathcal{X}^{T-D}(-m-1) & \xrightarrow[\cong_{\mu_d,B}]{\tilde{\varphi}} & \mathcal{Y}^{T-D}(-m-1) \\ \sigma^D \big\downarrow & & \sigma^D \big\downarrow \\ \mathcal{X}^T(-m) & \xrightarrow[\varphi]{\cong_{\mu_d,B}} & \mathcal{Y}^T(-m) \end{array} \tag{5.7}$$

and respecting the zero sections. Choose $n \geq 1$ such that $-(m+n) \equiv k \mod d$. For $s \gg 1$ after n iterations one arrives at an isomorphism $\mathcal{X}^{T-mD}(k) \cong_{\mu_d,B} \mathcal{Y}^{T-nD}(k)$. This holds for an arbitrary μ_d-stable divisor $T \in \operatorname{Div}(B)$. Replacing the initial T by $T+nD$ one gets an isomorphism $\mathcal{X}^T(k) \cong_{\mu_d,B} \mathcal{Y}^T(k)$, as required. □

In the proof we have used the following analog of Proposition 5.2.3. By abuse of notation we let v_i and $\tilde{v}_i$ be the local fiber coordinates of the line bundles $\mathcal{X}_l^T \to X$ and $\mathcal{Y}_l^T \to Y$, respectively.

PROPOSITION 5.4.11. *Under the assumptions of Theorem* 5.4.8 *let*

$$\psi_l\colon \mathcal{X}_l^T(-l) \overset{\cong_{\mu_d,B}}{\longrightarrow} \mathcal{Y}_l^T(-l)$$

be a μ_d-equivariant isomorphism such that

(i_l) $\psi_l^*(\tilde{v}_i) \equiv v_i \mod z^s \ \forall i$.

Then there exists a μ_d-equivariant isomorphism

$$\psi_{l+1}\colon \mathcal{X}_{l+1}^{T-D}(-l-1) \overset{\cong_{\mu_d,B}}{\longrightarrow} \mathcal{Y}_{l+1}^{T-D}(-l-1)$$

such that

(i_{l+1}) $\psi_{l+1}^*(\tilde{v}_i) \equiv v_i \mod z^{s-1} \ \forall i$.

HINT. The proof of Proposition 5.2.3 goes verbatim modulo the existence of an automorphism φ which is guaranteed by Theorem 4.1.4. Thus, it suffices to prove the following analog of Theorem 4.1.4.

THEOREM 5.4.12. *Let a GDF μ_d-surface $\pi_X\colon X \to B$, $z \in \mathcal{O}_B(B)$, and $T \in \operatorname{Div}(B)$ be as in Theorem* 5.4.8. *Then $\mathcal{X}^T$ satisfies an analog of the μ_d-equivariant condition $RF(l,-l,s)$.*

PROOF. It suffices to reproduce mutatis the proof of Theorem 4.1.4 (see Section 4.3). The modifications are as follows.

The coordinate v used when working with cylinders might do not exist on the total space of the line bundle $\pi^T\colon \mathcal{X}^T \to X$. Hence one cannot consider on $\mathcal{X}^T$ the locally nilpotent derivations $\tilde{\sigma}_{1,f}$ and $\tilde{\sigma}_{2,g}$ as in (4.5). However, one can use instead their analogs which coincide with these up to a given order on any special fiber component $\mathcal{F}_i = (\pi^T)^{-1}(F_i)$ in $\mathcal{X}^T$.

Indeed, let $\xi : L \to B$ be the line bundle associated with T, and let $U \subset B$ be a μ_d-stable dense open subset as in Lemma 5.4.5 which contains $z^{-1}(0) = \{b_1, \dots, b_n\}$ and such that $\xi|_U$ is trivial as a μ_d-line bundle. Then also the induced line bundle

$\pi^T\colon \mathcal{X}^T \to X$ is trivial over $V = \pi^{-1}(U) \subset X$. Thus, $\mathcal{X}^T|_V \cong_{\mu_d,V} V \times \mathbb{A}^1$ where $\mathbb{A}^1 = \operatorname{Spec} \Bbbk[v]$. Via this isomorphism, v yields a rational μ_d-quasi-invariant function on $\mathcal{X}^T$ which we denote by the same letter.

Choose a regular μ_d-quasi-invariant function $h \in \mathcal{O}_B(B)$ such that $h - 1 \equiv 0 \mod z^s$ and $h|_{B\setminus U} = 0$. Consider the lift $\tilde{h} \in \mathcal{O}_{\mathcal{X}^T}(\mathcal{X}^T)$ of h. For $s \gg 1$ the product $\tilde{v} = \tilde{h}^s v \in \mathcal{O}_{\mathcal{X}^T}(\mathcal{X}^T)$ is a regular μ_d-quasi-invariant which coincides with v to order s on any special fiber component $\mathcal{F}_i = (\pi^T)^{-1}(F_i)$ in $\mathcal{X}^T$. Letting $\partial_l^* = (\pi^T)^*(\partial_l)$ and $\hat{\sigma}_{1,f} = f(\tilde{v}^d)\partial_l^*$ for $f \in \Bbbk[t]$ yields a μ_d-invariant locally nilpotent derivation on $\mathcal{O}_{\mathcal{X}^T}(\mathcal{X}^T)$ which coincides to order s with $\tilde{\sigma}_{1,f}$ on any fiber component $\mathcal{F}_i$.

Furthermore, for $s \gg 1$ the product $\tilde{h}^{d+1}\partial/\partial v$ is a μ_d-invariant locally nilpotent derivation on $\mathcal{O}_{\mathcal{X}^T}(\mathcal{X}^T)$. Letting $\hat{\sigma}_{2,g} = \tilde{u}^{ds} g(\tilde{u}^d)\tilde{h}^d \partial/\partial v$ for $g \in \Bbbk[t]$ where $\tilde{u}$ is as defined in 4.3.1 yields a μ_d-invariant locally nilpotent derivation on $\mathcal{O}_{\mathcal{X}^T}(\mathcal{X}^T)$ which coincides to order s with $\tilde{\sigma}_{2,g}$ on any fiber component $\mathcal{F}_i$.

Using the locally nilpotent derivations $\hat{\sigma}_{1,f}$ and $\hat{\sigma}_{2,g}$ instead of $\tilde{\sigma}_{1,f}$ and $\tilde{\sigma}_{2,g}$, respectively, the rest of the proof of Theorem 4.1.4 applies and gives the desired μ_d-equivariant relative flexibility. □

CHAPTER 6

Basic examples of Zariski factors

6.1. Line bundles over affine curves

PROPOSITION 6.1.1. *Let $\pi: X \to B$ be a line bundle over a smooth affine curve B. Then the surface X is a Zariski factor.*

PROOF. If $B \cong \mathbb{A}^1$ then $\pi: X \to B$ is a trivial line bundle, and so, $X \cong \mathbb{A}^2$ is a Zariski factor by the Miyanishi-Sugie-Fujita Theorem ([**35**, **57**]; see also [**56**, Ch. 3, Thm. 2.3.1]).

Suppose further that $B \not\cong \mathbb{A}^1$, and so, any morphism $\mathbb{A}^1 \to B$ is constant. Consider a second smooth affine surface X' and the cylinders $\mathcal{X} = X \times \mathbb{A}^n$ and $\mathcal{X}' = X' \times \mathbb{A}^n$. Assume that there is an isomorphism $\varphi: \mathcal{X}' \xrightarrow{\cong} \mathcal{X}$. The structure of a vector bundle of $\tilde{\pi} := \mathrm{pr}_1 \circ \pi: \mathcal{X} \to B$ is transferred by φ^{-1} to such a structure on $\mathcal{X}'$ with the projection $\tilde{\pi}' = \tilde{\pi} \circ \varphi: \mathcal{X}' \to B$. This yields the commutative diagram

$$\begin{array}{ccc} \mathcal{X}' & \xrightarrow[\cong_B]{\varphi} & \mathcal{X} \\ {\scriptstyle \tilde{\pi}'}\downarrow & & \downarrow{\scriptstyle \tilde{\pi}} \\ B & \xrightarrow[\mathrm{id}]{} & B \end{array} \tag{6.1}$$

Since any morphism $\mathbb{A}^1 \to B$ is constant, $\tilde{\pi}'$ admits a factorization

$$\tilde{\pi}': \mathcal{X}' \xrightarrow{\mathrm{pr}_1} X' \xrightarrow{\pi'} B\,. \tag{6.2}$$

Letting $F_b = \pi^{-1}(b)$ and $F'_b = \pi'^{-1}(b) \subset X'$ where $b \in B$, φ restrits to an isomorphism

$$\varphi|_{F'_b \times \mathbb{A}^n}: F'_b \times \mathbb{A}^n \xrightarrow{\cong} F_b \times \mathbb{A}^n \cong \mathbb{A}^{n+1}\,.$$

Since any curve is a Zariski factor ([**1**, Thm. 6.5]) one deduces that $F'_b \cong F_b \cong \mathbb{A}^1$ $\forall b \in B$. Thus, $\pi': X' \longrightarrow B$ is an $\mathbb{A}^1$-fibration with all the fibers being reduced and irreducible, because the fibers of $\tilde{\pi}': \mathcal{X}' \to B$ are. Therefore, $\pi': X' \longrightarrow B$ admits a structure of a line bundle. Now the existence of an isomorphism $X' \cong X$ follows from the next lemma where we let $d = 1$. □

We work in an equivariant setup that we will use below.

LEMMA 6.1.2. *Consider two GDF μ_d-surfaces $\pi: X \to B$ and $\pi': X' \to B$ with only irreducible fibers. Extend the μ_d-actions to the cylinders $\mathcal{X} = X \times \mathbb{A}^n$ and $\mathcal{X}' = X' \times \mathbb{A}^n$ by the identity on the second factor. Suppose that there is a μ_d-equivariant isomorphism $\Phi: \mathcal{X} \xrightarrow{\cong_{\mu_d, B}} \mathcal{X}'$ over B. Then there exist μ_d-equivariant line bundle structures ξ and ξ' on X and X' with projections π and π', respectively, and a μ_d-equivariant isomorphism of line bundles $\xi \cong_{\mu_d, B} \xi'$ identical on B.*

PROOF. The average of an arbitrary section of π upon the μ_d-action on X yields a μ_d-invariant section of π and, respectively, a μ_d-equivariant structure of a line bundle ξ on X with projection π. Similarly, X' admits a μ_d-equivariant line bundle structure ξ' with projection π'. Let us show that there exists an isomorphism of line bundles $\xi \cong_{\mu_d,B} \xi'$.

The cylinder $\mathcal{X} = X \times \mathbb{A}^n$ inherits a structure of a vector bundle of rank $n+1$ isomorphic to the Whitney sum $\xi \oplus \mathbf{1_n}$ with projection $\tilde{\pi}: \mathcal{X} \to B$ where $\mathbf{1_n}$ stands for the trivial vector bundle of rank n over B. Similarly, $\tilde{\pi}': \mathcal{X}' \to B$ represents the vector bundle $\xi' \oplus \mathbf{1_n}$. The μ_d-equivariant B-isomorphism of the total spaces

$$\Phi: \mathcal{X} = \operatorname{tot}(\xi \oplus \mathbf{1_n}) \stackrel{\cong_{\mu_d,B}}{\longrightarrow} \mathcal{X}' = \operatorname{tot}(\xi' \oplus \mathbf{1_n})$$

sends the zero section Z of $\xi \oplus \mathbf{1_n}$ to a section, say, Z'' of $\xi \oplus \mathbf{1_n}$. Both Z and Z'' are μ_d-invariant. It is easily seen that the translation $t_{-Z''}$ on $-Z''$ in $\xi' \oplus \mathbf{1_n}$ is μ_d-equivariant. Hence the composition $\Psi = t_{-Z''} \circ \Phi$ is as well, and it sends Z to the zero section Z' of $\xi' \oplus \mathbf{1_n}$. The differential $d\Psi|_Z$ yields a μ_d-equivariant isomorphism of the normal bundles $\mathcal{N}_{Z/\mathcal{X}} \cong_{\mu_d,B} \mathcal{N}_{Z'/\mathcal{X}'}$. Furthermore, $\mathcal{N}_{Z/\mathcal{X}} \cong_{\mu_d,B} \xi \oplus \mathbf{1_n}$ and $\mathcal{N}_{Z'/\mathcal{X}'} \cong_{\mu_d,B} \xi \oplus \mathbf{1_n}$ as μ_d-vector bundles. Therefore, one has $\xi \oplus \mathbf{1_n} \cong_{\mu_d,B} \xi' \oplus \mathbf{1_n}$, that is, the line bundles ξ and ξ' are stably μ_d-equivariantly equivalent. In fact, they are μ_d-equivariantly equivalent. Indeed, one has (see [**66**, §8, Corollary])

$$\xi \cong_{\mu_d,B} \det(\xi \oplus \mathbf{1_n}) \cong_{\mu_d,B} \det(\xi' \oplus \mathbf{1_n}) \cong_{\mu_d,B} \xi',$$

as stated. □

6.2. Parabolic $\mathbb{G}_m$-surfaces: an overview

Let us start with the following auxiliary elementary lemma. For the sake of completeness we provide a brief argument.

LEMMA 6.2.1. *Let G be an algebraic group, let X be a G-variety, and let $\nu: X_{\text{norm}} \to X$ be the normalization morpism. Then X_{norm} admits a unique structure of G-variety making ν a G-equivariant morphism.*

PROOF. We claim the existence of a morphism Φ making the following diagram commutative:

$$\begin{array}{ccc} G \times X_{\text{norm}} & \stackrel{\Phi}{\longrightarrow} & X_{\text{norm}} \\ {\scriptstyle \text{id}\times\nu}\downarrow & \searrow^{\hat{\varphi}} & \downarrow{\scriptstyle \nu} \\ G \times X & \stackrel{\varphi}{\longrightarrow} & X \end{array}$$

where φ is the morphism of the G-action on X, and the diagonal arrow $\hat{\varphi}$ is just the composition of φ with $\text{id} \times \nu$. Indeed, the lift Φ of $\hat{\varphi}$ does exist due to the fact that $G \times X_{\text{norm}}$ and X_{norm} are both normal, and ν is the normalization morphism. Notice that the regular locus of X is G-invariant. Restricting the diagram over the regular loci of $G \times X$ and X shows that Φ satisfies the axioms of a G-action on the whole X_{norm} making the diagram G-equivariant. In particular, the morphism ν is G-equivariant. □

DEFINITIONS 6.2.2 (*the DPD presentation for parabolic $\mathbb{G}_m$-surfaces*). ([**28**]) A *parabolic $\mathbb{G}_m$-surface* is a normal affine surface X endowed with an effective $\mathbb{G}_m$-action along the fibers of an $\mathbb{A}^1$-fibration $\pi: X \to C$ over a smooth affine curve C.

The $\mathbb{G}_m$-action on X defines a grading

$$\mathcal{O}_X(X) = \bigoplus_{n\geq 0} A_n \quad \text{where} \quad A_n = H^0(C, \mathcal{O}_C(\lfloor nD_X \rfloor)\ \forall n \geq 0$$

for a $\mathbb{Q}$-divisor D_X on C. This is called a *Dolgachev-Pinkham-Demazure presentation*, or a *DPD presentation* for short, see [**28**, Thm. 3.2]. The $\mathbb{Q}$-divisor D_X on C is uniquely defined by the class of isomorphism of $\pi: X \to C$ up to a linear equivalence. Any fiber $\pi^*(p)$, $p \in C$, is irreducible of multiplicity m where $D_X(p) = (e/m)[p]$ with coprime $e, m \in \mathbb{Z}$. Any fiber $\pi^{-1}(p)$, taken with its reduced structure, is smooth and isomorphic to $\mathbb{A}^1$ ([**29**, Rem. 3.13(iii)]). The projection $\pi: X \to C$ admits a section consisting of the fixed points of the $\mathbb{G}_m$-action on X. The singularities of X are the fixed points of the $\mathbb{G}_m$-action in the multiple fibers of π. More precisely, if $D_X(p) = (e/m)[p]$ where $m > 1$ and e, m are coprime then the unique fixed point x_p over p is a cyclic quotient singularity of type (m, e') where $e' \in \{1, \ldots, m-1\}$ and $e' \equiv e \mod m$, see [**28**, Prop. 3.8(b)]. The following analog of Proposition 4.12 in [**28**] deals with parabolic (instead of hyperbolic) $\mathbb{G}_m$-surfaces.

LEMMA 6.2.3. *Given a parabolic $\mathbb{G}_m$-surface $\pi: X \to C$ and a branched covering $\mu: B \to C$, let $\pi': X' \to B$ be obtained from the cross-product $B \times_C X$ via normalization. Then $\pi': X' \to B$ is again a parabolic $\mathbb{G}_m$-surface. The $\mathbb{Q}$-divisors D_X on C and $D_{X'}$ on B in the corresponding DPD presentations are related via $D_{X'} = \mu^* D_X$.*

PROOF. The projection $\pi: X \to C$ is the orbit morphism of a parabolic $\mathbb{G}_m$-action, say, Λ on X with all its fibers being smooth and irreducible. Hence the fibers of $\pi': X' \to B$ are also irreducible and Λ lifts to a (parabolic) $\mathbb{G}_m$-action Λ' on the cross-product $B \times_C X$ where $\lambda: (b, x) \mapsto (b, \lambda.x)\ \forall \lambda \in \mathbb{G}_m$. By Lemma 6.2.1 this lifted action survives through the normalization $X' \to B \times_C X$. Thus, Λ lifts to a parabolic $\mathbb{G}_m$-action Λ' on X' such that $\pi': X' \to B$ is the orbit morphism. The induced morphism $\mu': X' \to X$ is $\mathbb{G}_m$-equivariant.

On the other hand, consider the $\mathbb{Q}$-divisor $D_{X''} = \mu^* D_X$ on B and the corresponding parabolic $\mathbb{G}_m$-surface $\pi'': X'' \to B$ with the DPD presentation related to the pair $(B, D_{X''})$. For any $n \geq 0$ there is a natural embedding

$$A_n = H^0(C, \mathcal{O}_C(\lfloor nD_X \rfloor)) \hookrightarrow \hat{A}_n = H^0(B, \mathcal{O}_B(\lfloor nD_{X''} \rfloor)).$$

This yields a monomorphism of graded rings

$$\mathcal{O}_X(X) = \bigoplus_{n\geq 0} A_n \hookrightarrow \mathcal{O}_{X''}(X'') = \bigoplus_{n\geq 0} A''_n$$

and the induced $\mathbb{G}_m$-equivariant surjection $\mu'': X'' \to X$ that fits in the commutative diagram

$$\begin{array}{ccc} X'' & \xrightarrow{\mu''} & X \\ {\scriptstyle \pi''}\downarrow & & \downarrow{\scriptstyle \pi} \\ B & \xrightarrow[\mu]{} & C \end{array}$$

By the universal property of the cross-product, μ'' can be factorized as

$$\mu'': X'' \to B \times_C X \xrightarrow{\pi} X.$$

Since X'' is normal one has as well a factorization $\mu'' : X'' \xrightarrow{\psi} X' \xrightarrow{\pi} X$ where ψ is a $\mathbb{G}_m$-equivariant surjection fitting in the commutative diagram

$$\begin{array}{ccc} X'' & \xrightarrow{\quad\psi\quad} & X' \\ & {\scriptstyle \tilde{\pi}''} \searrow \quad \swarrow {\scriptstyle \tilde{\pi}'} & \\ & B & \end{array} .$$

Since the fibers of a parabolic $\mathbb{G}_m$-surface are irreducible ([**29**, Rem. 3.13(iii)]), ψ is a bijection. Due to the normality of both X'' and X', ψ is an isomorphism. Now the desired conclusion follows. □

PROPOSITION 6.2.4. *Consider an $\mathbb{A}^1$-fibration $\pi : X \to C$ on a normal affine surface X over a smooth affine curve C. Let $\tilde{\pi} : \tilde{X} \to B$ be a marked GDF μ_d-surface obtained from $\pi : X \to C$ via a cyclic base change $\delta : B \to C$ with the Galois group μ_d and a subsequent normalization as in Lemma* 2.1.3. *Then the following are equivalent.*

(i) *$\tilde{\pi} : \tilde{X} \to B$ admits a structure of a line bundle;*
(ii) *$\pi : X \to C$ admits a structure of a parabolic $\mathbb{G}_m$-surface.*

PROOF. (i)⇒(ii). In case (i) the fibers of $\pi : X \to C$ are irreducible. If all of them are reduced then $\pi : X \to C$ admits a structure of a line bundle, and so, (ii) holds. Otherwise, $\pi : X \to C$ has multiple fibers. If a fiber $F_c = \pi^{-1}(c)$ is multiple then the branched covering construction of 2.1.2 creates a unique point $b = \delta^{-1}(c) \in B$ over c. This point b is a fixed point of the μ_d-action on B.

The μ_d-action on $\tilde{X}$ preserves the fibration $\tilde{\pi} : \tilde{X} \to B$ and sends the sections of $\tilde{\pi}$ to sections. Taking the fiberwise barycenter of the μ_d-orbit of the zero section yields a μ_d-invariant section, say, Z of $\tilde{\pi}$. There is a new line bundle structure on $\tilde{X}$ with projection $\tilde{\pi}$ and the associated parabolic $\mathbb{G}_m$-action $\tilde{\Lambda}$ on $\tilde{X}$ along the fibers of $\tilde{\pi}$ with Z as the fixed point set. In fact, $\tilde{\Lambda} = t_Z \circ \Lambda \circ t_Z^{-1}$ where Λ stands for the parabolic $\mathbb{G}_m$-action on $\tilde{X}$ associated with the original line bundle structure on $\tilde{X}$, and $t_Z \in \operatorname{Aut}_B \tilde{X}$ is the translation on Z in the vertical direction.

The $\mathbb{G}_m$-action $\tilde{\Lambda}$ commutes with the μ_d-action on $\tilde{X}$. Indeed, the conjugation of $\tilde{\Lambda}$ by elements of the μ_d-action yields a homomorphism $\mu_d \to \operatorname{Aut} \mathbb{G}_m \cong \mathbb{Z}/2\mathbb{Z}$. To show that this homomorphism is trivial we take a fixed point $b \in B$ of μ_d. The fiber $\tilde{F}_b = \tilde{\pi}^{-1}(b) \cong \mathbb{A}^1$ is μ_d-invariant and $Z \cap \tilde{F}_b$ is a common fixed point of μ_d and $\tilde{\Lambda}$. Hence $\mu_d|_{\tilde{F}_b} \subset \tilde{\Lambda}|_{\tilde{F}_b}$. Since μ_d and $\tilde{\Lambda}$ commute when restricted to $\tilde{F}_b$ they commute on $\tilde{X}$.

Therefore, $\tilde{\Lambda}$ descends to a parabolic $\mathbb{G}_m$-action on the quotient $X = \tilde{X}/\mu_d$ along the fibers of $\pi : X \to C$ thus converting X into a parabolic $\mathbb{G}_m$-surface over C.

(ii)⇒(i). Conversely, suppose that $\pi : X \to C$ is the orbit morphism of a parabolic $\mathbb{G}_m$-action Λ on X. Then Λ lifts to a parabolic $\mathbb{G}_m$-action $\tilde{\Lambda}$ on the cross-product $B \times_C X$ where $\lambda : (b, x) \mapsto (b, \lambda.x)$ $\forall \lambda \in \mathbb{G}_m$. By Lemma 6.2.1 this lifted action survives through the normalization $\tilde{X} \to B \times_C X$. Thus, Λ lifts through the branched covering $\tilde{X} \to X$ as in 2.1.2. In this way the GDF surface $\tilde{\pi} : \tilde{X} \to B$ acquires an effective parabolic $\mathbb{G}_m$-action $\tilde{\Lambda}$ along the fibers of $\tilde{\pi}$. Hence all these fibers are reduced and irreducible, cf. [**29**, Rem. 3.13(iii)]. This allows to define a line bundle structure on $\tilde{\pi} : \tilde{X} \to B$. □

REMARKS 6.2.5. 1. The integral divisor $-\mathcal{D}_{\tilde{X}} = -\mu^* D_X \in \operatorname{Pic} B$ is associated with the line bundle $\tilde{\pi}: \tilde{X} \to B$. This divisor $\mathcal{D}_{\tilde{X}}$ determines a DPD presentation of the $\mathbb{G}_m$-surface $\tilde{\pi}: \tilde{X} \to B$ (see Lemma 6.2.3).
2. It is known that a Gizatullin $\mathbb{G}_m$-surface X is toric if and only if the associated extended graph Γ_{ext} is linear, see [**30**, Lem. 2.20]. A similar criterion holds for the parabolic $\mathbb{G}_m$-surfaces. Namely, one can show that conditions (i) and (ii) of Proposition 6.2.4 are equivalent to the following one (cf. [**30**, Prop. 3.22]):
(iii) Γ_{ext} *is a bush, that is, any fiber tree* $\Gamma_c(\bar{\pi})$, $c \in C$, *is a chain.*

As usual, Γ_{ext} stands for the extended graph of a pseudominimal resolved completion $\bar{\pi}: \bar{X} \to \bar{C}$ of the $\mathbb{A}^1$-fibered surface $\pi: X \to C$ as in Proposition 6.2.4. This graph is viewed as a rooted tree with the section at infinity S as the root vertex.

6.3. Parabolic $\mathbb{G}_m$-surfaces as Zariski factors

The following theorem is the main result of Section 6.

THEOREM 6.3.1. *Any parabolic $\mathbb{G}_m$-surface is a Zariski factor.*

The proof is done in Lemmas 6.3.6–6.3.15. It is preceded by several auxiliary facts. In the next elementary lemma we use the following terminology.

DEFINITION 6.3.2. Given a morphism $\pi: X \to C$ of a normal affine variety X onto a smooth curve C, the fiber over a point $c \in C$ is called *multiple* if $d > 1$ where d is the greatest common divisor of the multiplicities of the components of $\pi^*(c)$.

LEMMA 6.3.3. *A polynomial of one variable cannot have two or more multiple fibers.*

PROOF. Suppose that $f \in \Bbbk[t]$ (is nonconstant and) has at least two multiple fibers, say, $f^*(0)$ and $f^*(1)$. Then $f = p^r = 1 - q^s$ for some polynomials $p, q \in \Bbbk[t]$ such that $p^r + q^s = 1$, $r, s \geq 2$, $\deg p = d/r$, and $\deg q = d/s$ where $d = \deg f$. The derivative f' vanishes to order $r-1$ at any root of p and to order $s-1$ at any root of q. More precisely, since p and q do not have any common root one has

$$\operatorname{div} f' \geq (r-1) \operatorname{div} p + (s-1) \operatorname{div} q\,.$$

Taking the degrees one gets the inequalities

$$(r-1)/r + (s-1)/s \leq (d-1)/d\,.$$

Since $r, s \geq 2$ it follows that

$$1 \leq \left(1 - \frac{1}{r}\right) + \left(1 - \frac{1}{s}\right) \leq \left(1 - \frac{1}{d}\right).$$

This gives a contradiction. Alternatively, letting $x = p(t)$, $y = q(t)$ yields a parametrization of the plane affine curve $E = \{x^r + y^s = 1\}$. However, there is no nonconstant morphism $\mathbb{A}^1 \to E$. □

An affine variety is called $\mathbb{A}^1$-*uniruled* if a general point of X belongs to the image of a nonconstant morphism $\mathbb{A}^1 \to X$. One can find in the literature different versions of the following results, see, e.g., [**64**] and [**42**, Thm. 4.1].

LEMMA 6.3.4. *Consider a dominant morphism $\pi: X \to C$ from an affine variety X to a smooth affine curve C. Assume that one of the following conditions is fulfilled.*

(i) $C \not\cong \mathbb{A}^1$;
(ii) $C \cong \mathbb{A}^1$ *and π has at least two multiple fibers.*

Then the following hold.

(a) *Any morphism* $\mathbb{A}^n \to X$ *has image contained in a fiber of* π. *Consequently, there is no dominant morphism* $\mathbb{A}^n \to X$.

(b) *If the general fibers of* π *are* $\mathbb{A}^1$*-uniruled then any automorphism* $\alpha \in \operatorname{Aut} X$ *preserves the fibration* $\pi: X \to C$, *that is, sends the fibers to fibers.*

PROOF. (a) In case (i) any morphism $\mathbb{A}^n \to C$ is constant, hence the assertion follows. Assuming (ii) suppose to the contrary that there exists $F: \mathbb{A}^n \to X$ such that $f := \pi \circ F: \mathbb{A}^n \to \mathbb{A}^1$ is nonconstant. Then $f \in \Bbbk[t_1, \dots, t_n]$ is a nonconstant polynomial with two distinct multiple fibers. The latter contradicts Lemma 6.3.3.

(b) If $\alpha \in \operatorname{Aut} X$ does not preserve the fibration $\pi: X \to C$ then there is a morphism $\varphi: \mathbb{A}^1 \to X$ such that the composition $f = \pi \circ \varphi$ is not constant. Then $C \cong \mathbb{A}^1$, and so, $f \in \Bbbk[t]$ is a polynomial with two multiple fibers. This leads to a contradiction as before. □

In the proof of Theorem 6.3.1 we use the following auxiliary lemma.

LEMMA 6.3.5. *Let* $\pi: X \to \mathbb{P}^1$ *be an* $\mathbb{A}^1$*-fibration on a normal affine surface* X. *Assume that the group* $\operatorname{Pic} X$ *is finite and* $\pi(X) \supset \mathbb{A}^1 = \mathbb{P}^1 \setminus \{\infty\}$. *Then* $\pi(X) = \mathbb{A}^1$, *all fibers of* π *are irreducible, and the divisor class group* $\operatorname{Cl}(X)$ *is generated by the classes of the multiple fibers of* π.

PROOF. Let $\bar{\pi}: \bar{X} \to \mathbb{P}^1$ be a resolved completion of $\pi: X \to \mathbb{P}^1$ with extended graph Γ_{ext}. Then Γ_{ext} is a rooted tree with the section at infinity S as the root. We let

- $\mathcal{B}_1, \dots, \mathcal{B}_n$ be the degenerate fibers of $\bar{\pi}$ over the points $b_i \in \mathbb{P}^1$, $i = 1, \dots, n$;
- $D = \bar{X} \setminus X_{\text{resolved}}$ stand for the boundary divisor;
- E stand for the exceptional divisor of the resolution of singularities $X_{\text{res}} \to X$;
- $m_i \geq 0$ be the number of components of the fiber $\pi^{-1}(b_i)$;
- $n_i \geq 0$ be the number of components of $\mathcal{B}_i$ which are components of $D+E$.

Thus, $\mathcal{B}_i$ consists of $n_i + m_i$ components. Contracting subsequently (-1)-fiber components one arrives finally at a Hirzebruch surface $\mathbb{F}_s$. In this way one contracts $n_i + m_i - 1$ components of $\mathcal{B}_i$, $i = 1, \dots, n$. Let $\varrho(V)$ be the Picard number of a variety V. Since $\varrho(\mathbb{F}_s) = 2$ one has $\varrho(\bar{X}) = 2 + \sum_{i=1}^n (n_i + m_i - 1)$. Letting $\natural \mathcal{D}$ be the number of components of a divisor $\mathcal{D}$ one gets

$$0 = \varrho(X) = \varrho(\bar{X}) - \natural(D+E)$$

$$= \left(2 + \sum_{i=1}^n (n_i + m_i - 1)\right) - \left(1 + \sum_{i=1}^n n_i\right) = 1 + \sum_{i=1}^n (m_i - 1). \tag{6.3}$$

It follows that

- $m_i \leq 1 \ \forall i = 1, \dots, n$, that is, the fibers of $\pi: X \to \pi(X)$ are irreducible;
- $m_i = 0$ for exactly one value of i, that is, $\pi(X) = \mathbb{A}^1$.

Let $\omega \subset \mathbb{A}^1$ be a Zariski open dense subset such that $U = \pi^{-1}(\omega)$ is isomorphic over ω to the cylinder $\omega \times \mathbb{A}^1$, and so, $\operatorname{Cl}(U) = 0$. For $\mathcal{D} = X \setminus U$ one has the exact sequence $\operatorname{Div}(\mathcal{D}) \to \operatorname{Cl}(X) \to \operatorname{Cl}(U) \to 0$ where $\operatorname{Div}(\mathcal{D})$ is the subgroup of Weil divisors on X supported by $\mathcal{D}$, see [**56**, p. 206]. Thus $\operatorname{Cl}(X)$ is generated by the fibers of π contained in $\mathcal{D}$. Any reduced fiber of π represents the zero class in $\operatorname{Cl}(X)$. Hence $\operatorname{Cl}(X)$ is generated by the classes of the multiple fibers of π. □

The next lemma gives a proof of Theorem 6.3.1 under an additional assumption.

LEMMA 6.3.6. *Let $\pi: X \to C$ be a parabolic $\mathbb{G}_m$-surface, and let X' be a normal affine surface. Assume that there is an isomorphism $\varphi: \mathcal{X}' \xrightarrow{\cong} \mathcal{X}$ of the n-cylinders $\mathcal{X} = X \times \mathbb{A}^n$ and $\mathcal{X}' = X' \times \mathbb{A}^n$. Suppose also that for the induced $\mathbb{A}^{n+1}$-fibration $\hat{\pi}: \mathcal{X} \to C$ one of the conditions* (i) *and* (ii) *of Lemma 6.3.4 is fulfilled. Then $X' \cong X$.*

PROOF. By a *ruling* of $\mathcal{X}$ we mean a fiber $\{a\} \times \mathbb{A}^n$ of the first projection $\mathcal{X} \to X$ where $a \in X$. By Lemma 6.3.4 the composition $\hat{\pi} \circ \varphi$ restricts to a constant map of any ruling $\{x'\} \times \mathbb{A}^n$ to C where $x' \in X'$. This provides a surjection $\pi': X' \to C$ which extends to a morphism $\hat{\pi}' = \pi' \circ \mathrm{pr}_1: \mathcal{X}' \to C$ fitting in the commutative diagram

$$\begin{array}{ccc} \mathcal{X}' & \xrightarrow[\cong]{\varphi} & \mathcal{X} \\ {\scriptstyle \hat{\pi}'}\downarrow & & \downarrow{\scriptstyle \hat{\pi}} \\ C & \xrightarrow[\psi]{\cong} & C \end{array} \tag{6.4}$$

For any point $c \in C$, φ restricts to an isomorphism

$$\varphi|_{\pi'^{-1}(c) \times \mathbb{A}^n}: \pi'^{-1}(c) \times \mathbb{A}^n \xrightarrow{\cong} \pi^{-1}(\psi(c)) \times \mathbb{A}^n \cong \mathbb{A}^{n+1}\,.$$

Since any curve is a Zariski factor ([**1**, Thm. 6.5]) and each fiber of π is isomorphic to $\mathbb{A}^1$ one deduces that

- any fiber of $\pi': X' \to C$ is isomorphic to $\mathbb{A}^1$;
- ψ induces a correspondence between the multiple fibers of π' (and $\hat{\pi}'$) and the ones of π (and $\hat{\pi}$, respectively) preserving the multiplicities, and each of these multiple fibers carries a unique singular point of X', see 6.2.2.

Applying to the surfaces $\pi: X \to C$ and $\pi': X' \to C$ the branched covering constructions with suitable cyclic base changes $\mu: B \to C$ and $\mu': B' \to C$ as in Lemma 2.1.3 one obtains two GDF μ_d-surfaces $\tilde{\pi}: \tilde{X} \to B$ and $\tilde{\pi}': \tilde{X}' \to B'$ and two cyclic branched coverings $\tilde{X} \to X$ and $\tilde{X}' \to X'$ with the Galois group μ_d. The coverings μ and μ' can be chosen fitting in the commutative diagram

$$\begin{array}{ccc} B' & \xrightarrow[\cong_{\mu_d}]{\tilde{\psi}} & B \\ {\scriptstyle \mu'}\downarrow & & \downarrow{\scriptstyle \mu} \\ C & \xrightarrow[\psi]{\cong} & C \end{array}$$

The same branched covering constructions applied to the cylinders $\mathcal{X}$ and $\mathcal{X}'$ (which are isomorphic over C up to an automorphism of C, see (6.4)) yields the cylinders $\tilde{\mathcal{X}} = \tilde{X} \times \mathbb{A}^n$ and $\tilde{\mathcal{X}}' = \tilde{X}' \times \mathbb{A}^n$ fitting in a μ_d-equivariant commutative diagram

$$\begin{array}{ccc} \tilde{\mathcal{X}}' & \xrightarrow[\cong_{\mu_d}]{\tilde{\varphi}} & \tilde{\mathcal{X}} \\ {\scriptstyle \hat{\tilde{\pi}}'}\downarrow & & \downarrow{\scriptstyle \hat{\tilde{\pi}}} \\ B' & \xrightarrow[\tilde{\psi}]{\cong_{\mu_d}} & B \end{array} \tag{6.5}$$

where $\tilde{\varphi}$ is a lift of φ from (6.4), and $\tilde{\psi}$ is a lift of ψ.

By Proposition 6.2.4, $\tilde{\pi}: \tilde{X} \to B$ admits a line bundle structure. In particular, the fibers of $\tilde{\pi}: \tilde{X} \to B$ and the ones of $\hat{\tilde{\pi}}: \hat{\mathcal{X}} \to B$ are reduced and irreducible. Hence the fibers of $\hat{\tilde{\pi}}': \tilde{\mathcal{X}}' \to B$ and the ones of $\tilde{\pi}': \tilde{X}' \to B$ are as well. Therefore, $\tilde{\pi}': \tilde{X}' \to B$ admits as well a structure of a line bundle. Proceeding as in the proof of (i)⇒(ii) in Proposition 6.2.4 one can choose μ_d-equivariant line bundle structures, say, ξ and ξ' of $\tilde{\pi}: \tilde{X} \to B$ and $\tilde{\pi}': \tilde{X}' \to B'$, respectively. Then the zero sections of ξ and ξ' are μ_d-invariant. Taking the quotients by the μ_d-actions yields a structure of parabolic $\mathbb{G}_m$-surfaces on $\pi: X \to C$ and $\pi': X' \to C$ where the former is the given $\mathbb{G}_m$-structure. By Lemma 6.1.2 there is a μ_d-equivariant isomorphism of line bundles $\xi' \cong \xi$ inducing the isomorphism $\tilde{\psi}: B' \xrightarrow{\cong_{\mu_d}} B$. It induces a $\mathbb{G}_m$-equivariant isomorphism of the quotient parabolic $\mathbb{G}_m$-surfaces $X \cong_{\mathbb{G}_m} X'$ which is a lift of the automorphism ψ of C. □

6.3.7. We will suppose in the sequel that in the setting of Theorem 6.3.1 neither (i) nor (ii) of Lemma 6.3.4 is fulfilled, that is, $C = \mathbb{A}^1$ and the fibration $\pi: X \to \mathbb{A}^1$ has at most one multiple fiber. By virtue of the Miyanishi-Sugie-Fujita Theorem we may suppose as well that the parabolic $\mathbb{G}_m$-surface $\pi: X \to \mathbb{A}^1$ has exactly one multiple fiber $\pi^{-1}(0)$ of multiplicity $d > 0$, and so, X has precisely one singular point, say, x, which is the unique fixed point of the $\mathbb{G}_m$-action on the multiple fiber $\pi^{-1}(0)$ and a cyclic quotient singularity. Due to the following lemma, one can recover this singularity from the isomorphism type of a cylinder $\mathcal{X} = X \times \mathbb{A}^n$. Indeed, let X' be a normal affine surface such that the cylinders $\mathcal{X}$ and $\mathcal{X}'$ are isomorphic. Then by Lemma 6.3.8, X' has as well precisely one singular point, say, x', and there is a local isomorphism of singularities $(X, x) \cong (X', x')$.

LEMMA 6.3.8. *Let $\pi: X \to C$ be a normal affine surface with a unique singular point $x \in X$, and let X' be a normal affine surface. Suppose there is an isomorphism $\varphi: \mathcal{X}' \xrightarrow{\cong} \mathcal{X}$ of the n-cylinders $\mathcal{X} = X \times \mathbb{A}^n$ and $\mathcal{X}' = X' \times \mathbb{A}^n$. Then X' has as well a unique singular point, say, x', and the germs of surface singularities (X, x) and (X', x') are isomorphic.*

PROOF. The singular locus of $\mathcal{X}$ consists of the unique ruling $\{x\} \times \mathbb{A}^n$, and the one of $\mathcal{X}'$ is a disjoint union of the rulings over the singular points of X'. Since $\varphi: \mathcal{X}' \xrightarrow{\cong} \mathcal{X}$ sends $\operatorname{sing} \mathcal{X}' = (\operatorname{sing} X') \times \mathbb{A}^n$ isomorphically onto $\operatorname{sing} \mathcal{X} = \{x\} \times \mathbb{A}^n$, then X' also has a unique singular point x', and $\varphi(\{x'\} \times \mathbb{A}^n) = \{x\} \times \mathbb{A}^n$.

Consider further the morphism $\sigma_1: X_1 \to X$ which consists in blowing up the maximal ideal of the unique singular point $x \in X$ followed by the normalization. The induced morphism of n-cylinders $\sigma_1 \times \mathrm{id}_{\mathbb{A}^n}: \mathcal{X}_1 \to \mathcal{X}$ consists in blowing up the ideal of the singular locus $\operatorname{sing} \mathcal{X} = \{x\} \times \mathbb{A}^n$ and the subsequent normalization. By a theorem of Zariski ([**72**]; see also [**51**]) a sequence of blowups in maximal ideals and subsequent normalizations

$$X_N \xrightarrow{\sigma_N} X_{N-1} \to \ldots \to X_1 \xrightarrow{\sigma_1} X_0 = X$$

resolves the singularity (X, x). It induces a similar sequence of blowups in rulings of our n-cylinders and subsequent normalizations

$$\mathcal{X}_N \xrightarrow{\sigma_N} \mathcal{X}_{N-1} \to \ldots \to \mathcal{X}_1 \xrightarrow{\sigma_1} \mathcal{X}_0 = \mathcal{X}$$

which results in a resolution of the corresponding singularities of $\mathcal{X} \cong \mathcal{X}'$. The exceptional divisor of the resolution $\mathcal{X}_N \to \mathcal{X}$ is $\mathcal{E} = E \times \mathbb{A}^n$ where E is the exceptional divisor of the resolution $X_N \to X$.

Let further $\sigma_1': X_1' \to X'$ be the blowup of the maximal ideal of the singular point $x' \in X'$ followed by the normalization. Then $\sigma_1' \times \mathrm{id}_{\mathbb{A}^n}: \mathcal{X}_1' \to \mathcal{X}'$ is the blowup of the ideal sheaf of the singular locus $\operatorname{sing} \mathcal{X}' = \{x'\} \times \mathbb{A}^n$ followed by the normalization. Under the isomorphism $\varphi: \mathcal{X}' \xrightarrow{\cong} \mathcal{X}$ the ruling $\{x'\} \times \mathbb{A}^n = \operatorname{sing} \mathcal{X}'$ goes to the ruling $\{x\} \times \mathbb{A}^n = \operatorname{sing} \mathcal{X}$ inducing an isomorphism of the corresponding ideal sheaves and of the blowups in these ideal sheaves. By an argument used in the proof of Lemma 6.2.1, the latter isomorphism lifts to an isomorphism of the normalizations. Finally, φ lifts to an isomorphism $\varphi_1: \mathcal{X}_1' \xrightarrow{\cong} \mathcal{X}_1$.

Continuing in this way one arrives finally at a resolution $\mathcal{X}_N' \to \mathcal{X}'$ where $\mathcal{X}_N' = X_N' \times \mathbb{A}^n \cong \mathcal{X}_N$ with exceptional divisor $\mathcal{E}' = E' \times \mathbb{A}^n \cong \mathcal{E} = E \times \mathbb{A}^n$ where E' is the exceptional divisor of the induced resolution of singularity $X_N' \to X'$. Under this procedure the singularities of the embedded surfaces $X \times \{0\} \subset \mathcal{X}$ and $X' \times \{0\} \subset \mathcal{X}'$ are simultaneously resolved, and there is an isomorphism $\varphi_N: \mathcal{X}_N' \xrightarrow{\cong} \mathcal{X}_N$ such that $\varphi_N(\mathcal{E}') = \mathcal{E}$. The only irreducible complete curves in $\mathcal{E}$ (in $\mathcal{E}'$, respectively) are of the form $E_i \times \{v\}$ ($E_i' \times \{v'\}$, respectively) where E_i and E_i' are components of E and E', respectively, and $v, v' \in \mathbb{A}^n$. Given such a curve $E_i' \times \{v'\}$ there is a curve $E_{\sigma(i)} \times \{v\}$ such that $\varphi_N(E_i' \times \{v'\}) = E_{\sigma(i)} \times \{v\}$. It follows that $\varphi_N(E_i' \times \mathbb{A}^n) = E_{\sigma(i)} \times \mathbb{A}^n$. The image $\varphi_N(X_N' \times \{v'\})$ is a smooth surface in $\mathcal{X}_N$ which meets the exceptional divisor $\mathcal{E} \subset \mathcal{X}_N$ transversely along the curve $\varphi_N(E' \times \{v'\}) = E \times \{v\} \subset X_N \times \{v\}$. The same is true for $X_N \times \{v\}$. Namely, the latter is a smooth surface in $\mathcal{X}_N$ which meets $\mathcal{E}$ transversely along the same curve $E \times \{v\}$. Projecting the both surfaces to X_N via the canonical projection $\mathcal{X}_N \to X_N$ yields a local isomorphism of the surface germs $(\varphi_N(X_N' \times \{v'\}), E \times \{v\})$ and $(X_N \times \{v\}, E \times \{v\})$ near the common exceptional divisor $E \times \{v\}$. Contracting the divisor $E \times \{v\}$ yields an isomorphism between the singular surface germs (X, x) and (X', x'). □

6.3.9. Remind that a toric variety X is called *non-degenerate* if $\mathcal{O}_X(X)^\times = \mathbb{k}^*$. Any non-degenerate affine toric surface is isomorphic to the quotient $V_{d,e}$ of $\mathbb{A}^2$ by a diagonal μ_d-action

$$(6.6) \qquad \zeta.(x,y) = (\zeta x, \zeta^e y) \quad \text{where} \quad \zeta^d = 1, \ \ 1 \le e < d, \ \ \gcd(e,d) = 1$$

(see, e.g., [**28**, Ex. 2.3] or [**29**, Ex. 2.8]). One has $\mathrm{Cl}(\mathbb{A}^2/\mu_d) \cong \mathbb{Z}/d\mathbb{Z}$. Moreover, $V_{d,e} \cong V_{d',e'}$ if and only if $d = d'$ and either $e = e'$, or $ee' \equiv 1 \mod d$, see [**28**, Ex. 2.3]. The same is true for the cyclic surface singularities $(V_{d,e}, v)$ and $(V_{d',e'}, v')$, where v and v' are the respective images of the origin of $\mathbb{A}^2$. It follows that $V_{d,e} \cong V_{d',e'}$ if and only if $(V_{d,e}, v) \cong (V_{d',e'}, v')$.

LEMMA 6.3.10. *Under the assumptions of* 6.3.7, *X is a non-degenerate affine toric surface.*

PROOF. According to Proposition 6.2.4 the branched covering construction applied to $\pi: X \to \mathbb{A}^1$ with the cyclic base change $\mathbb{A}^1 \to \mathbb{A}^1$, $z \mapsto z^d$, yields a a line bundle $\tilde{\pi}: \tilde{X} \to \mathbb{A}^1$ which is trivial since $\operatorname{Pic} \mathbb{A}^1 = 0$. Its zero section is μ_d-invariant, hence the line bundle structure is μ_d-equivariant. Via an isomorphism $\tilde{X} \cong \mathbb{A}^2$ one obtains an effective action of μ_d on $\mathbb{A}^2$ which can be linearized taking the form (6.6) in appropriate coordinates on $\mathbb{A}^2$. Then $\tilde{\pi}$ becomes the standard projection $\mathbb{A}^2 \to \mathbb{A}^1$, $(x,y) \mapsto x$. Thus $X \cong \mathbb{A}^2/\mu_d$ is a non-degenerate affine toric surface of type (d,e). □

COROLLARY 6.3.11. *Under the assumptions of* 6.3.7 *one has* $\mathrm{Cl}(X') \cong \mathrm{Cl}(X) \cong \mathbb{Z}/d\mathbb{Z}$.

PROOF. Recall (see [**34**, Thm. 8.1]; cf. [**35**, (9.9.8)]) that $\mathrm{Cl}(X)$ is a cancellation invariant. Since $\mathcal{X} \cong \mathcal{X}'$ there are isomorphisms

$$\mathrm{Cl}(X') \cong \mathrm{Cl}(\mathcal{X}') \cong \mathrm{Cl}(\mathcal{X}) \cong \mathrm{Cl}(X) \cong \mathbb{Z}/d\mathbb{Z}\,. \tag{6.7}$$

□

We use below the following simple version of the Cox ring (see [**3**], [**15**]).

DEFINITION 6.3.12 (*Cox ring*). Let X be a normal affine variety with $\mathcal{O}_X(X)^\times = \Bbbk^*$. Suppose that the divisor class group $\mathrm{Cl}(X)$ is a finite cyclic group of order d generated by the class of a Weil divisor F on X. Consider the $(\mathbb{Z}/d\mathbb{Z})$-graded Cox ring

$$\mathrm{Cox}\,\mathcal{O}_X(X) := \bigoplus_{j=0}^{d-1} H^0(X, \mathcal{O}_X(jF))\zeta^j$$

where $\zeta \in k^\times$ is a primitive dth root of unity. Then $\tilde{X} = \mathrm{Spec}\,\mathrm{Cox}\,\mathcal{O}_X(X)$ is a normal affine variety equipped with a μ_d-action defined via the $(\mathbb{Z}/d\mathbb{Z})$-grading on $\mathcal{O}_{\tilde{X}}(\tilde{X}) = \mathrm{Cox}\,\mathcal{O}_X(X)$, see [**3**, Thm. 1.5.1.1]. The natural embedding $\mathcal{O}_X(X) \hookrightarrow \mathrm{Cox}\,\mathcal{O}_X(X)$ onto the subalgebra of μ_d-invariants yields the quotient morphism $\tilde{X} \to X = \tilde{X}/\mu_d$. We call this morphism a *Cox covering construction.*

Consider the n-cylinder $\mathcal{X} = X \times \mathbb{A}^n$ over X. The divisor class group $\mathrm{Cl}\,\mathcal{X} \cong \mathrm{Cl}(X) \cong \mu_d$ is generated by the class of the Weil divisor $\mathcal{F} = F \times \mathbb{A}^n$ on $\mathcal{X}$, see [**34**, Thm. 8.1]. The Cox covering construction applied to $(\mathcal{X}, \mathcal{F})$ yields the n-cylinder $\tilde{\mathcal{X}} = \tilde{X} \times \mathbb{A}^n$.

For the next lemma we provide two alternative proofs.

LEMMA 6.3.13. *In the setup of* 6.3.7*, X' is a non-degenerate affine toric surface.*

PROOF. By Lemma 6.3.10 one has $\mathcal{O}_X(X)^\times = \Bbbk^*$. Since $\mathcal{O}_{\mathcal{X}'}(\mathcal{X}')^\times \cong \mathcal{O}_{\mathcal{X}}(\mathcal{X})^\times = \Bbbk^*$ then also $\mathcal{O}_{X'}(X') = \Bbbk^*$. Consider the Weil divisor $F_0 = \pi^{-1}(0)$ on X. Its class generates the class group $\mathrm{Cl}(X) \cong \mu_d$. By [**4**, Thm. 3.1] the branched covering $\mathbb{A}^2 \to X = \mathbb{A}^2/\mu_d$ as in Lemma 6.3.10 coincides with the Cox covering defined by the pair (X, F_0), see Definition 6.3.12. Letting $\mathcal{F}_0 = F_0 \times \mathbb{A}^n$ and applying the cyclic Cox covering construction to the pair $(\mathcal{X}, \mathcal{F}_0)$ one obtains the n-cylinder $\tilde{\mathcal{X}} = \mathbb{A}^2 \times \mathbb{A}^n = \mathbb{A}^{n+2}$.

Choose a Weil divisor F_0' on X' whose class in $\mathrm{Cl}(X')$ is sent to the class of F_0 in $\mathrm{Cl}(X)$ via the isomorphisms (6.7). Applying the Cox covering construction to the pair (X', F_0') leads to a cyclic μ_d-covering $\tilde{X}' \to X'$. Letting $\mathcal{F}_0' = F_0' \times \mathbb{A}^n$ and applying the Cox covering construction to the pair $(\mathcal{X}', \mathcal{F}_0')$ yields the n-cylinder $\tilde{\mathcal{X}}' = \tilde{X}' \times \mathbb{A}^n$. We claim that $\tilde{\mathcal{X}}'$ is isomorphic to $\tilde{\mathcal{X}} \cong \mathbb{A}^{n+2}$. Indeed, let $\varphi\colon \mathcal{X} \xrightarrow{\cong} \mathcal{X}'$ be an isomorphism. One has $\mathcal{F}_0' \sim \varphi_*\mathcal{F}_0$ on $\mathcal{X}'$. Up to an isomorphism, the Cox covering construction does not depend on the choice of a divisor in the class generating the group $\mathrm{Cl}(\mathcal{X}') \cong \mathbb{Z}/d\mathbb{Z}$, see [**3**, Prop. 1.4.2.2]. Hence applying this construction to the pair $(\mathcal{X}', \varphi_*\mathcal{F}_0)$ yields a variety, say, $\hat{\mathcal{X}}'$ isomorphic to $\tilde{\mathcal{X}}'$. On the other hand, the isomorphism of pairs $\varphi\colon (\mathcal{X}, \mathcal{F}_0) \xrightarrow{\cong} (\mathcal{X}', \varphi_*\mathcal{F}_0)$ leads to an isomorphism $\hat{\mathcal{X}}' \cong \tilde{\mathcal{X}}$.

It follows that $\tilde{X}' \times \mathbb{A}^n \cong \mathbb{A}^{n+2}$. By the Miyanishi-Sugie-Fujita Theorem ([**35**, Cor. 3.3], [**56**, Ch. 3, Thm. 2.3.1]) one has $\tilde{X}' \cong \mathbb{A}^2$. Since the μ_d-action on $\mathbb{A}^2$ can be linearized (see, e.g., [**38**, Thm. 2]) the quotient $X' \cong \mathbb{A}^2/\mu_d$ is an affine toric surface. □

The following alternative proof of Lemma 6.3.13 avoids to address the Cox rings.

6.3.14 (*The second proof of Lemma* 6.3.13). Alternatively, one can argue as follows.

CLAIM 1. There exists an $\mathbb{A}^1$-fibration $\pi' : X' \to \mathbb{A}^1$.

PROOF OF CLAIM 1. Due to the Iitaka-Fujita Theorem ([**44**]) the log-Kodaira dimension is a cancellation invariant. Hence $\bar{k}(X') = \bar{k}(X) = -\infty$. Therefore, X' admits an $\mathbb{A}^1$-fibration $\pi' : X' \to C$, see [**56**, Ch. 2, Thm. 2.1.1]. Composing the induced surjection $\mathcal{X}' \to C$ with an isomorphism $\varphi : \mathcal{X} \xrightarrow{\cong} \mathcal{X}'$ one concludes that either $C \cong \mathbb{A}^1$ or $C \cong \mathbb{P}^1$. However, the latter is impossible. Indeed, by Corollary 6.3.11 one has $\mathrm{Cl}(X') \cong \mathrm{Cl}(X) \cong \mathbb{Z}/d\mathbb{Z}$ where d is the multiplicity of the fiber $F_0 = \pi^{-1}(0)$ through the singular point $x \in X$. Let F_0' be the fiber of π' through the singular point x' of X'. Since $\mathrm{Cl}(X') \cong \mathbb{Z}/d\mathbb{Z}$ the divisor dF_0' on X' is principal, that is, $dF_0' = \mathrm{div}\, f$ for some $f \in \mathcal{O}_{X'}(X')$. Since f is constant on any $\mathbb{A}^1$-fiber of π' one has $f = \pi'^*(g)$ where $g \in \mathcal{O}_C(C)$ is not a constant. Hence C cannot be a complete curve.

CLAIM 2. Let $\pi' : X' \to \mathbb{A}^1$ be an $\mathbb{A}^1$-fibration. Then each fiber of π' is irreducible, there exists exactly one multiple fiber of π', this fiber has multiplicity d and contains the singular point x' of X'.

PROOF OF CLAIM 2. The irreducibility of the fibers of π' follows from Lemma 6.3.5. Suppose to the contrary that $\pi' : X' \to \mathbb{A}^1$ has two or more multiple fibers. Then by Lemma 6.3.4 any automorphism of the cylinder $\mathcal{X}'$ preserves the induced $\mathbb{A}^{n+1}$-fibration $\mathcal{X}' \to \mathbb{A}^1$. The same must be true for $\mathcal{X} \cong \mathcal{X}'$ and the induced fibration $\mathcal{X} \to \mathbb{A}^1$. However, the cylinder $\mathcal{X}$ over the non-degenerate affine toric surface X is flexible ([**5**, Thm. 2.1]). This leads to a contradiction.

Let $F_0' = (\pi')^{-1}(0)$ be the unique multiple fiber of π'. By Corollary 6.3.11 one has

$$\mathrm{Cl}\,(X') = \langle F_0' \rangle \cong \mathrm{Cl}\,(X) = \langle F_0 \rangle \cong \mathbb{Z}/d\mathbb{Z}\,. \tag{6.8}$$

It follows that $(\pi')^*(0) = dF_0'$ and, in turn, $\pi'(x') = 0 \in \mathbb{A}^1$. This proves Claim 2.

Consider further pseudominimal resolved completions $\bar{\pi} : \bar{X} \to \mathbb{P}^1$ and $\bar{\pi}' : \bar{X}' \to \mathbb{P}^1$ of $\pi : X \to \mathbb{A}^1$ and $\pi' : X' \to \mathbb{A}^1$, respectively. Let $\bar{F}_0 \subset \bar{X}$ and $\bar{F}_0' \subset \bar{X}'$ be the fiber components which contain the proper transforms of F_0 and F_0', respectively. Since the fibers $F_0 = \pi^{-1}(0)$ and $F_0' = (\pi')^{-1}(0)$ are irreducible and the completions are pseudominimal, $\bar{F}_0$ and $\bar{F}_0'$ are the unique (-1)-vertices of $\Gamma_0(\bar{\pi})$ and $\Gamma_0(\bar{\pi}')$, respectively. These are the bridges of the unique feathers $\mathcal{F} \subset \Gamma_0(\bar{\pi})$ and $\mathcal{F}' \subset \Gamma_0(\bar{\pi}')$, respectively. The chains $\mathcal{F} \ominus \bar{F}_0$ and $\mathcal{F}' \ominus \bar{F}_0'$ correspond to the exceptional divisors of the minimal resolutions of the cyclic quotient singularities (X, x) and (X', x'), respectively. By Lemma 6.3.8 one has $(X, x) \cong (X', x')$. Hence $\mathcal{F} \cong \mathcal{F}'$ as ordered chains.

For the non-degenerate affine toric surface X the extended graph Γ_{ext} of $(\bar{X}, D)$ is a chain ([**30**, Lem. 2.20]). Hence also the fiber tree $\Gamma_0(\bar{\pi})$ is. The tip, say, R of the chain $\Gamma_0(\bar{\pi})$ which is not a tip of $\mathcal{F}$ meets the section at infinity S. Hence R has multiplicity 1 in $\bar{\pi}^*(0)$. It follows that the chain $\mathcal{B} = \Gamma_0(\bar{\pi}) \ominus R$ can be contracted to a smooth point starting with the contraction of the unique (-1)-component $\bar{F}_0$, see Lemma 2.3.2.

Claim 3. The extended graph Γ'_{ext} of $(\bar{X}', D')$ is a chain.

Proof of Claim 3. It suffices to show that the fiber tree $\Gamma_0(\bar{\pi}')$ is a chain. Suppose the contrary. Then the extremal linear branch $\mathcal{B}'$ of $\Gamma_0(\bar{\pi}')$ which contains $\mathcal{F}'$ is adjacent to a branching vertex, say, R' of $\Gamma_0(\bar{\pi}')$. The contraction of $\mathcal{F}'$ makes R' a (-1)-vertex of degree 2 which contracts to a node of the resulting fiber. Hence R' has multiplicity $m > 1$ in $(\pi')^*(0)$.

As follows from [**30**, Lem. 4.7] the isomorphism $\mathcal{F} \cong \mathcal{F}'$ can be extended to an isomorphism of the contractible chains $\mathcal{B} \supset \mathcal{F}$ and $\mathcal{B}' \supset \mathcal{F}'$ with the unique (-1)-vertices $\bar{F}_0$ and $\bar{F}'_0$, respectively. Since the multiplicity of $\bar{F}_0$ in $\pi^*(0)$ equals d the one of $\bar{F}_0$ in $(\pi')^*(0)$ equals $md > d$. However, the latter contradicts (6.8). This proves the claim.

Since the extended graph Γ'_{ext} is a chain and $\mathcal{O}_{X'}(X')^\times = \Bbbk^\times$ it follows that X' is a Gizatullin surface. Applying [**30**, Lem. 2.20] one concludes that X' is a nondegenerate affine toric surface. □

The next lemma completes the proof of Theorem 6.3.1.

Lemma 6.3.15. *Under the assumptions of* 6.3.7 *one has* $X' \cong X$.

Proof. Two non-degenerate toric affine surfaces are isomorphic if and only if their singularities are, see 6.3.9. By Lemmas 6.3.10 and 6.3.13, X and X' are non-degenerate toric affine surfaces, and by Lemma 6.3.8 the singularities (X, x) and (X', x') are isomorphic. Hence $X \cong X'$. □

CHAPTER 7

Zariski 1-factors

7.1. Stretching and rigidity of cylinders

Let us introduce the following combinatorial transformations of graph divisors $\mathcal{D}(\pi)$ (see Definition 2.3.11).

DEFINITION 7.1.1 (*Combinatorial stretchings*). Let B be a smooth affine curve. Given an effective divisor $A = \sum_{i=1}^n a_i p_i \in \operatorname{Div}(B)$ where $a_i \in \mathbb{Z}_{\geq 0}$ and $p_i \in B$, $i = 1, \ldots, n$ we associate with A a chain divisor $\mathcal{D}(A) = \sum_{i=1}^n L(a_i) p_i$ where $L(a_i)$ is a chain with weights $[[-2, -2, \ldots, -2, -1]]$ of length a_i if $a_i > 0$ and $L(0) = \varnothing$ otherwise.

Let $\mathcal{D} = \sum_i \Gamma_i p_i$ be a graph divisor where Γ_i is a rooted tree with a root v_i. The *combinatorial top level A-stretching* is the transformation

$$\mathcal{D} \mapsto A.\mathcal{D} = \mathcal{D}' = \sum_{i=1}^n \Gamma_i' p_i$$

where Γ_i' is obtained from Γ_i by inserting the chain L_i above each vertex v of Γ_i on the top level $\operatorname{ht}(\Gamma_i)$ so that the (-1)-vertex of L_i becomes a tip of Γ_i'; the weight of v decreases by 1.

A combinatorial A-stretching is called *principal* if $A = \operatorname{div} f$ where $f \in \mathcal{O}_B(B) \setminus \{0\}$.

The *combinatorial* $(A, \overline{-1})$*-stretching* consists in inserting the chain L_i of length a_i below the root v_i of Γ_i, so that the (-1)-vertex r_i of L_i becomes the root vertex of the new rooted tree Γ_i'.

DEFINITION 7.1.2 (*Geometric stretching*). Let $\pi' \colon X' \to B$ and $\pi \colon X \to B$ be two GDF surfaces over B, and let again $A = \sum_{i=1}^n a_i p_i$ where $a_i \geq 0$ $\forall i$ and $\{p_1, \ldots, p_n\} = \operatorname{supp} \mathcal{D}(\pi)$. An affine modification $\sigma : X' \to X$ over B will be called a *geometric A-stretching* if its effect on the graph divisor $\mathcal{D}(\pi)$ amounts to a combinatorial A-stretching, that is, if $\mathcal{D}(\pi') = A.\mathcal{D}(\pi)$ is as in Definition 7.1.1.

A *geometric* $(A, \overline{-1})$*-stretching* inserts the chain $[[-2, \ldots, -2, -1]]$ of length a_i in the fiber tree $\Gamma_{p_i}(\pi)$ between the root and the section S so that the (-1)-vertex of this chain becomes the root of the resulting fiber tree $\Gamma_{p_i}(\pi')$. A *principal geometric* $(A, \overline{-1})$*-stretching* with $A = \operatorname{div} f$ for $f \in \mathcal{O}_B(B) \setminus \{0\}$ amounts to perform in (2.7) an affine modification $X_0' \to X_0$ of $X_0 = B \times \mathbb{A}^1$ with the divisor $(f \circ \pi)^*(0)$ and the center $f^*(0) \times \{0\}$. In other words, letting $\mathbb{A}^1 = \operatorname{Spec} \Bbbk[u]$ one has

$$\mathcal{O}_{X_0'}(X_0') = \mathcal{O}_B(B)[u'] \quad \text{where} \quad u' = u/f\,. \tag{7.1}$$

Therefore, $X_0' \cong_B B \times \mathbb{A}^1$. Performing the remaining fibered modifications in (2.7) gives again the same surface $X = X_m$.

Let us supply an example of a geometric stretching.

EXAMPLE 7.1.3. Following Example 3.5.1 consider the nth Danielewski surface $\pi_n: X_n \to \mathbb{A}^1$, where X_n is given in $\mathbb{A}^3$ with coordinates (z, u, t_n) by equation $z^n t_n - u^2 + 1 = 0$, and the projection $\pi_n: X_n \to B = \mathbb{A}^1$ is given by $(z, u, t_n) \mapsto z$. The function $t_n = t_0/z^n$ yields (modulo z) a natural affine coordinate on each component of the special fiber $z = 0$. The morphism $\varrho_n: X_n \to X_{n-1}$ is given by

$$\varrho_n: (z, u, t_n) \mapsto (z, u, t_{n-1} = zt_n) .$$

For any $k \in \{1, \dots, n-1\}$ the morphism

$$\varrho_{k+1} \circ \dots \circ \varrho_n: X_n \to X_k, \qquad (z, u, t_n) \mapsto (z, u, t_k = z^{n-k} t_n) ,$$

amounts to a principal top level geometric A-stretching with $A = \operatorname{div} z^{n-k}$. This stretching replaces the fiber tree $\mathcal{D}(\pi_k) = \Gamma_0(\pi_k)\cdot[0]$ by $\mathcal{D}(\pi_n) = \Gamma_0(\pi_n)\cdot[0]$ where (cf. (3.1))

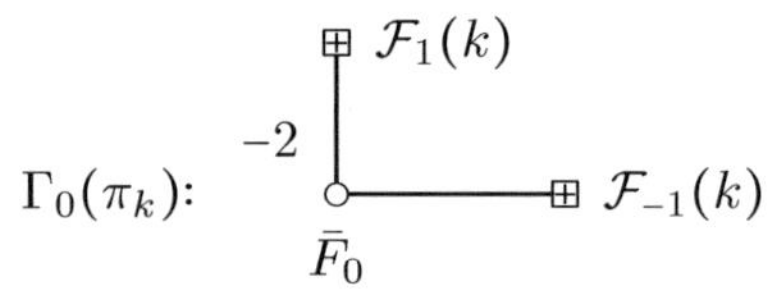

(7.2)

with $\mathcal{F}_i(k) = [[-2, \dots, -2, -1]]$, $i \in \{1, -1\}$ being the chains of length k. For instance, the principal top level A-stretching $X_4 \to X_2$ with $A = \operatorname{div} z^2$ replaces the fiber tree $\Gamma_0(\pi_2)$ by $\Gamma_0(\pi_4)$, see Figure 7.1.

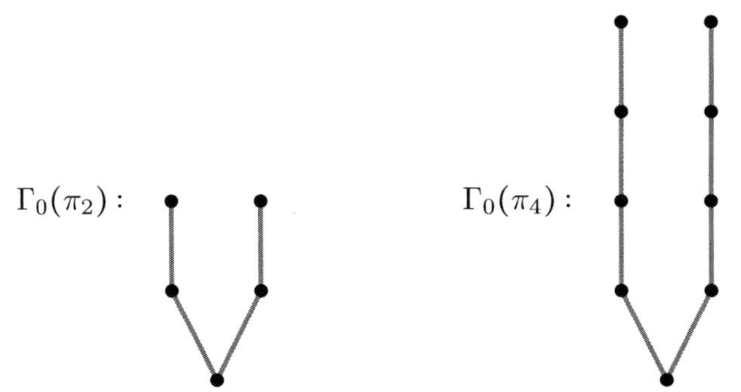

FIGURE 7.1. A principal top level stretching of $\Gamma_0(\pi_2)$ yields $\Gamma_0(\pi_4)$

In the following lemma we provide a simple geometric realization of certain combinatorial stretchings more general than the top level ones.

LEMMA 7.1.4. *Let $\pi: X \to B$ be a marked GDF surface with a marking $z \in \mathcal{O}_B(B) \setminus \{0\}$ and a graph divisor $\mathcal{D}(\pi) = \sum_{i=1}^n \Gamma_i b_i$, and let $\mathfrak{F}$ be the set of leaves of $\mathcal{D}(\pi)$. Given a subset $\mathfrak{F}_0$ of $\mathfrak{F}$ and an integer $s \gg 1$ choose a function $\tilde{u} \in \mathcal{O}_X(X)$ such that*

(i) *$\tilde{u} \equiv u_F \mod z^s$ near any fiber component F with $\bar{F} \in \mathfrak{F}_0$ and*
(ii) *$\tilde{u} \equiv z^s \mod z^{s+1}$ near any fiber component F with $\bar{F} \in \mathfrak{F} \setminus \mathfrak{F}_0$.*

Choose also a function $f \in \mathcal{O}_B(B)$ such that $\operatorname{supp}(\operatorname{div} f) \subset \operatorname{supp}(\operatorname{div} z)$. *Letting*

$$X' = \operatorname{Spec} \mathcal{O}_X(X)[\tilde{u}/f]$$

consider the morphisms $\pi': X' \to B$ and $\sigma: X' \to_B X$ associated to the natural embeddings $\mathcal{O}_B(B) \subset \mathcal{O}_X(X) \subset \mathcal{O}_{X'}(X')$. Then $\pi': X' \to B$ is a GDF surface with *the graph divisor* $\mathcal{D}(\pi') = \sum_{i=1}^{n} \Gamma'_i b_i$ *obtained from* $\mathcal{D}(\pi)$ *by attaching to each leaf* $\bar{F} \in \mathfrak{F}_0$ *with* $\pi(F) = b_i$ *a chain* $L_i = [[-2, \dots, -2, -1]]$ *of length* $a_i := \mathrm{ord}_{b_i}(f)$.

PROOF. For a component F of $\pi^{-1}(b_i)$ one has $(f \circ \pi)|_{U_F} \equiv c_i z^{a_i} \mod z^{a_i+1}$ where $c_i \neq 0$. Due to (ii) if $\bar{F} \in \mathfrak{F} \setminus \mathfrak{F}_0$ then the morphism $\sigma: X' \to X$ is an isomorphism over U_F.

If $\bar{F} \in \mathfrak{F}_0$ then $\sigma: X' \to X$ restricted to $\sigma^{-1}(U_F)$ is an a_i-iterated fibered modification with a reduced divisor F and center at the maximal ideal (z, u_F) and its infinitesimally near points. Indeed, let $P_0 = \{z = 0, u_F = 0\}$ be the origin of the local coordinate system (z, u_F) in U_F near F. The affine modification of X along $z = 0$ with center P_0 amounts in U_F to the extension $\mathcal{O}_{U_F}(U_F) \hookrightarrow \mathcal{O}_{U_F}(U_F)[u_F/z]$. Iterating one obtains the extension $\mathcal{O}_{U_F}(U_F) \hookrightarrow \mathcal{O}_{U_F}(U_F)[u_F/z^{a_i}]$. Due to (i) the latter extension coincides with $\mathcal{O}_{U_F}(U_F)[\tilde{u}/f]$. Hence the fiber tree Γ'_i is obtained from Γ_i by joining the left end of the chain $L_i = [[-2, \dots, -2, -1]]$ of length a_i to the leaf $\bar{F}$ of Γ_i. By contrast, for $\bar{F} \in \mathfrak{F} \setminus \mathfrak{F}_0$ one has $\mathcal{O}_{U_F}(U_F) = \mathcal{O}_{U_F}(U_F)[\tilde{u}/z]$. □

REMARKS 7.1.5. 1. Changing the order of choices in Lemma 7.1.4, one can start with arbitrary principal effective divisor $A = \mathrm{div}\, f$ where $f \in \mathcal{O}_B(B)$ and then choose a new marking $z \in \mathcal{O}_B(B)$ for X such that $\mathrm{supp}(\mathrm{div}\, f) \subseteq \mathrm{supp}(\mathrm{div}\, z)$. If $\mathfrak{F}_0 = \mathfrak{F}$ in Lemma 7.1.4 and every leaf $\bar{F} \in \mathfrak{F}$ with $\pi(F) = b_i$ is on the same (top) level $\mathrm{ht}(\Gamma_i)$, $i = 1, \dots, n$ then $\sigma: X' \to X$ in this lemma is the principal top-level A-stretching.

2. Let $\pi': X' \to B$ and $\pi'': X'' \to B$ be two marked GDF μ_d-surfaces equivariantly dominating X over B with the same graph divisors $\mathcal{D}(\pi') = \mathcal{D}(\pi'')$. Then by Theorem 5.2.2 there is a μ_d-equivariant isomorphism of cylinders $\mathcal{X}'(k) \cong_{\mu_d, B} \mathcal{X}''(k)$ $\forall k \in \mathbb{Z}$. Thus, different geometric realizations of the same combinatorial stretching give rise to isomorphic cylinders.

EXAMPLE 7.1.6. Let us apply a stretching as in Lemma 7.1.4 to the Danielewski surface X_2. Choosing for $\mathcal{F}_0$ in Lemma 7.1.4 one of the two leaves of the graph $\Gamma_0(\pi_2)$ one can produce, e.g., an $\mathbb{A}^1$-fibered surface $\pi': X' \to A^1$ with the fiber tree $\Gamma_0(\pi')$ as in Figure 7.2 which does not correspond to any Danielewski surface.

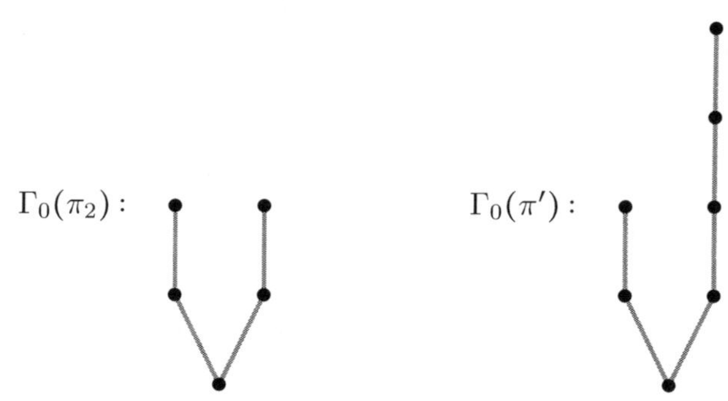

FIGURE 7.2. A stretching of the fiber tree $\Gamma_0(\pi_2)$ as in Lemma 7.1.4

COROLLARY 7.1.7. *Under the assumptions of Lemma* 7.1.4 *assume that* $\pi: X \to B$ *is a marked GDF* μ_d*-surface, and let* $\mathfrak{F}_0 \subset \mathfrak{F}$ *be a* μ_d*-invariant set of top level leaves of* $\mathcal{D}(\pi)$. *Then there exists a marked GDF* μ_d*-surface* $\pi': X' \to B$ *and an* *equivariant affine modification $\sigma: X' \to X$ over B which amounts to attaching a chain $L_F = [[-2,\ldots,-2,-1]]$ of the same length $a \geq 1$ to every leaf $\bar{F} \in \mathfrak{F}_0$ of the graph divisor $\mathcal{D}(\pi)$.*

PROOF. By Corollary 3.4.2 one can choose a μ_d-quasi-invariant function $\tilde{u}$ of weight $-l$ as in Lemma 7.1.4. Applying this lemma with $f = z^a$, the resulting surface $\pi': X' \to B$ where $X' = \operatorname{Spec} \mathcal{O}_X(X)[\tilde{u}/f]$ is a marked GDF μ_d-surface, the fibered modification $\sigma: X' \to X$ is μ_d-equivariant, and its effect on the graph divisors $\mathcal{D}(\pi)$ and $\mathcal{D}(\pi')$ is as desired. □

To describe the geometric realization of the combinatorial (-1)-stretching we need to introduce the following basic notions.

DEFINITION 7.1.8 (*The Danielewski-Fieseler quotient*). Given a GDF surface $\pi: X \to B$ the *Danielewski-Fieseler quotient* $\mathrm{DF}(\pi)$ is the quotient of X by the equivalence relation defined by the fiber components of π. Thus, $\mathrm{DF}(\pi)$ is a (non-separated, in general) one-dimensional scheme, and π factorizes as follows:

$$\pi: X \xrightarrow{p} \mathrm{DF}(\pi) \xrightarrow{q} B$$

where the fibers of $p: X \to \mathrm{DF}(\pi)$ are reduced and irreducible. In particular, $q: \mathrm{DF}(\pi) \to B$ is an isomorphism over $B \setminus \{b_1,\ldots,b_n\}$, while the total transform of b_i in $\mathrm{DF}(\pi)$ consists of N_i points $(b_{i,j})_{j=1,\ldots,N_i}$ where N_i is the number of the fiber components $F_{i,j}$ in $\pi^{-1}(b_i)$. Therefore, q is an isomorphism if and only if all fibers of π are irreducible, if and only if $\pi: X \to B$ admits a structure of a line bundle.

DEFINITION 7.1.9 (*Type divisors*). Let $l_{i,j} = l(F_{i,j})$ be the level of $F_{i,j}$. The anti-effective divisor on $\mathrm{DF}(\pi)$

$$\mathrm{tp}.\mathrm{div}(\pi) = -\sum_{i,j} l_{i,j} b_{i,j}$$

is called the *type divisor* of X.

The linear equivalence of divisors on $\mathrm{DF}(\pi)$ is, as usual, the equivalence modulo the principal divisors. Since the morphism $q: \mathrm{DF}(\pi) \to B$ is birational, any rational function on $\mathrm{DF}(\pi)$ is the lift of some rational function on B. Hence the principal divisors on $\mathrm{DF}(\pi)$ are the lifts of the ones on B via q. The Picard group $\mathrm{Pic}(\mathrm{DF}(\pi))$ is the group of classes of divisors on $\mathrm{DF}(\pi)$ modulo the linear equivalence.

EXAMPLE 7.1.10. The Danielewski-Fieseler quotient $\mathrm{DF}(\pi_n)$ of the Danielewski surface $\pi_n: X_n \to B = \mathbb{A}^1$ (see Examples 3.5.1 and 7.1.3) is the affine line with double origin $0_+, 0_-$ over $0 \in \mathbb{A}^1$. It is obtained from two copies of $\mathbb{A}^1$ by gluing them together along $\mathbb{A}^1_* = \mathbb{A}^1 \setminus \{0\}$. The morphism $q: \mathrm{DF}(\pi_n) \to \mathbb{A}^1$ is the tautological projection identifying 0_+ and 0_- with the origin $0 \in \mathbb{A}^1$. Thus, $\mathrm{DF}(\pi_n)$ does not depend on $n \geq 1$. By contrast, the type divisor $\mathrm{tp.div}(\pi_n) = -n([0_+] + [0_-])$ does depend on n. However, this is a principal divisor because $\mathrm{tp.div}(\pi_n) = q^*(-n[0])$ is the pullback of a principal divisor on $\mathbb{A}^1$. In particular, the class of $\mathrm{tp.div}(\pi_n)$ in the Picard group $\mathrm{Pic}(\mathrm{DF}(\pi_n))$ is zero, and so, any two such type divisors are linearly equivalent. Due to Theorem 0.0.6, the cylinders $\mathcal{X}_n$, $n = 1, 2, \ldots$ are pairwise isomorphic over $B = \mathbb{A}^1$.

The following lemma contains the promised application of the (-1)-stretching.

LEMMA 7.1.11. *Consider two GDF surfaces $\pi_X: X \to B$ and $\pi_Y: Y \to B$ with the same Danielewski-Fieseler quotient and with linearly equivalent type divisors,* *see Definition* 7.1.9. *Then one can choose new trivializing sequences* (2.7) *for* X *and* Y *in such a way that the corresponding type divisors coincide.*

PROOF. The principal geometric $(A,\overline{-1})$-stretching with $A = \operatorname{div} f \geq 0$ applied to a GDF-surface $\pi: X \to B$ affects the completion $(\hat{X}, \hat{D})$ as in (2.8) in such a way that

$$\text{(7.3)} \qquad \mathrm{tp.div}\,(\pi') = \mathrm{tp.div}\,(\pi) - q^*(A)\,.$$

In other words, the type divisor $\mathrm{tp.div}(\pi_X)$ is defined only up to adding a principal anti-effective divisor.

Since under our assumptions, the type divisors $\mathrm{tp.div}(\pi_X)$ and $\mathrm{tp.div}(\pi_Y)$ are linearly equivalent, there is a principal divisor $T \in \operatorname{Div} B$ such that

$$\text{(7.4)} \qquad \mathrm{tp.div}(\pi_X) = \mathrm{tp.div}(\pi_Y) + q^*(T)\,.$$

Let $A \in \operatorname{Div} B$ be a principal effective divisor such that $A - T \geq 0$. Performing the principal $(A,\overline{-1})$-stretching and $(A-T,\overline{-1})$-stretching, respectively, one replaces the trivializing sequences (2.7) for X and Y by suitable new ones so that the new type divisors for X and Y are

$$\mathrm{tp.div}(\pi_X) - q^*(A), \quad \text{resp.,} \quad \mathrm{tp.div}(\pi_Y) - q^*(A-T)\,,$$

see (7.3). Due to (7.4) the latter divisors are equal. □

As another application of the principal $(A,\overline{-1})$-stretching we have the following lemma.

LEMMA 7.1.12. *Consider a marked GDF surface* $\pi: X \to B$ *with a marking* $z \in \mathcal{O}_B(B)$ *where* $z^*(0) = b_1 + \ldots + b_n$, *a trivializing sequence* (2.7), *and the corresponding graph divisor* $\mathcal{D}(\pi) = \sum_{i=1}^n \Gamma_i b_i$. *Then the following hold.*

(a) *Performing a suitable principal geometric* $(A,\overline{-1})$*-stretching and extending* (2.7) *accordingly one may assume that*

$$\text{(7.5)} \qquad \mathrm{ht}(\Gamma_i) = \mathrm{ht}(\mathcal{D}(\pi)) \qquad \forall i = 1,\ldots,n\,.$$

(b) *Let, furthermore,* $\pi: X \to B$ *be a marked GDF* μ_d*-surface. Then* (7.5) *holds after performing a suitable principal* μ_d*-equivariant* $(A,\overline{-1})$*-stretching where* $A = \operatorname{div} f$ *with a* μ_d*-invariant function* $f \in \mathcal{O}_B(B) \setminus \{0\}$.

PROOF. (a) Let

$$D_\infty = \sum_{c_i \in \bar{B} \setminus B} c_i \quad \text{and} \quad D_0 = \sum_{i=1}^n (m - m_i) b_i \quad \text{where} \quad m = \mathrm{ht}\,(\mathcal{D}(\pi)) \quad \text{and} \quad m_i = \mathrm{ht}(\Gamma_i)\,.$$

Let also $D(r) = rD_\infty - D_0$ where $r \gg 1$. The very ample linear system $|D(r)|$ defines an embedding

$$\text{(7.6)} \qquad \Phi_{|D(r)|}: \bar{B} \hookrightarrow \mathbb{P}^{N(r)} = \mathbb{P}E \quad \text{where} \quad E = H^0(\bar{B}, \mathcal{O}_{\bar{B}}(D(r)))^\vee\,.$$

A general hyperplane section cuts out on $\bar{B}$ a reduced effective divisor, say, A' such that $\operatorname{supp} A' \subset B \setminus \operatorname{supp} z^*(0)$. Letting $A = A' + D_0 \sim rD_\infty$, where $\sim$ stands for the linear equivalence, there exists a rational function f on $\bar{B}$ with $\operatorname{div} f = A - rD_\infty \in \operatorname{Div} \bar{B}$. Then $A = \operatorname{div}(f|_B)$ is a principal effective divisor on B. Applying the $(A,\overline{-1})$-stretching over X one replaces the type divisor $\mathrm{tp.div}\,(\pi_X)$ by the new one

$$\mathrm{tp.div}\,(\pi_X) - q^*(A) = \mathrm{tp.div}\,(\pi_X) - q^*(A') - q^*(D_0)\,,$$

see (7.3). Due to our choice of A' and D_0, the new fiber tree, say, $\mathcal{D}'(\pi)$ satisfies (7.5) with $\mathcal{D}(\pi)$ replaced by $\mathcal{D}'(\pi)$.

(b) Under the assumptions of (b) the divisors D_0, D_∞, and $D(r)$ are μ_d-invariant and the embedding (7.6) is equivariant with respect to a linear action of μ_d on E. Consider the character decomposition $E = \bigoplus_{\chi\in\mu_d^\vee} E_\chi$. Choose $\chi \in \mu_d^\vee$ such that the equivariant projection $E \to E_\chi$ yields a non-constant map $\bar{B} \to \mathbb{P}E_\chi$. The reduced μ_d-invariant effective divisor A' cut out on $\bar{B}$ by a general hyperplane section in $\mathbb{P}E_\chi$ still satisfies $\operatorname{supp} A' \subset B \smallsetminus \operatorname{supp} z^*(0)$. Define $f|_B \in \mathcal{O}_B(B)$ as before. The divisor $A = \operatorname{div}(f|_B) \in \operatorname{Div}(B)$ being μ_d-invariant, f is a quasi-invariant of weight, say, $k \in \{0, \ldots, d-1\}$. Replacing f by the μ_d-invariant $\tilde{f} = z^{d-k} f$ one obtains a μ_d-invariant principal divisor

$$\tilde{A} = \operatorname{div}(\tilde{f}|_B) = A' + D_0' + (d-k)\operatorname{div} z^*(0)\,.$$

Then $\tilde{\mathcal{D}}(\pi) = (\tilde{A}.\mathcal{D}(\pi))_{\overline{-1}}$ verifies (7.5) with $\tilde{\mathcal{D}}(\pi)$ instead of $\mathcal{D}(\pi)$. □

REMARK 7.1.13. Extending a trivializing sequence (2.7) as in the proof one introduces new special fibers. The resulting graph divisor $\tilde{\mathcal{D}}(\pi)$ adopts a certain number of new fiber graphs $\tilde{\Gamma}_{p_j}(\pi) = [[0]]p_j$ of height 1 supported off $\operatorname{supp}(\operatorname{div} z) = \{b_1, \ldots, b_n\}$. So, the former marking z cannot serve any longer as a marking, and should be replaced by a suitable new one, which does exist, of course.

As we observed already in Definition 7.1.2, the principal $(A, \overline{-1})$-stretching does not affect the GDF surface $X \to B$. Due to the next proposition, a principal top-level A-stretching does not affect the cylinder $\mathcal{X}$ up to an isomorphism over B.

PROPOSITION 7.1.14. *Let $\pi: X \to B$ be a marked GDF surface with a marking $z \in \mathcal{O}_B(B)$ where* $\operatorname{div} z = b_1 + \ldots + b_n$. *Suppose that*

(α) *for $i = 1, \ldots, n$ the leaves of $\Gamma_{b_i}(\pi)$ are on the same level m_i.*

Let $\sigma: X' \to X$ be a principal top-level A-stretching as in Definition 7.1.2 *where $A = \operatorname{div} f$ for some $f \in \mathcal{O}_B(B) \smallsetminus \{0\}$. Then for any $s \gg 1$ there is an isomorphism of cylinders $\varphi: \mathcal{X} \xrightarrow{\cong_B} \mathcal{X}'$ such that for every pair of special fiber components F in X and F' in X' with $\varphi(F \times \mathbb{A}^1) = F' \times \mathbb{A}^1$ one has*[1]

$$(\varphi|_{U_F\times\mathbb{A}^1})_*: (z, u_F, v) \mapsto (z, u_{F'}, v') \mod z^s \tag{7.7}$$

in suitable natural coordinates (z, u_F, v) and $(z, u_{F'}, v')$ in the standard affine charts $U_F \times \mathbb{A}^1 \subset \mathcal{X}$ and $U_{F'} \times \mathbb{A}^1 \subset \mathcal{X}'$, respectively.

PROOF. (a) Choosing a new marking $z \in \mathcal{O}_B(B)$ for X one may suppose that $\operatorname{supp}(\operatorname{div} f) \subset \operatorname{supp}(\operatorname{div} z)$ and (α) still holds. Performing a suitable principal $(A, \overline{-1})$-stretching as in Lemma 7.1.12(a) one may assume that

(α_0) *for all $i = 1, \ldots, n$ the leaves of $\Gamma_{b_i}(\pi)$ are on the top level* $\operatorname{ht}(\mathcal{D}(\pi))$.

Consider the Asanuma modification of the second kind $\kappa: \mathcal{X}'' \to \mathcal{X}$ associated with f (see Definition 5.1.2), that is,

$$\mathcal{O}_\mathcal{X}(\mathcal{X}) \subset \mathcal{O}_{\mathcal{X}''}(\mathcal{X}'') = \mathcal{O}_X(X)[v/f]\,.$$

Lemma 5.1.4(a) provides an isomorphism

$$\beta: \mathcal{X} \xrightarrow{\cong_B} \mathcal{X}'' \quad \text{where} \quad \beta_*: (z, u_F, v) \mapsto (z, u_{F''}, v'') \tag{7.8}$$

[1]We let $\varphi_*(f, g, h) = (\varphi_* f, \varphi_* g, \varphi_* h)$ where $\varphi_* = (\varphi^{-1})^*$.

in suitable natural local coordinates. We claim that there is an isomorphism $\mathcal{X}' \cong_B \mathcal{X}''$. Indeed, due to Remark 7.1.5.2 and condition (α_0) one may suppose that

$$\mathcal{O}_{\mathcal{X}'}(\mathcal{X}') \cong_{\Bbbk[z]} \mathcal{O}_{\mathcal{X}}(\mathcal{X})[\tilde{u}/f]$$

where $\tilde{u} \in \mathcal{O}_X(X)$ verifies conditions (i) and (ii) of Lemma 7.1.4 with $\mathfrak{F}_0$ being the set of all top level fiber components in X. Thus, it suffices to establish an isomorphism

$$\mathcal{O}_{\mathcal{X}}(\mathcal{X})[\tilde{u}/f] \cong_{\mathcal{O}_B(B)} \mathcal{O}_{\mathcal{X}}(\mathcal{X})[v/f]. \tag{7.9}$$

By Lemma 4.4.2 there exists $\tau \in \mathrm{SAut}_B\, \mathcal{X}$ such that

$$(\tau|_{U_F \times \mathbb{A}^1})_*: (z, u_F, v) \mapsto (z, v, -u_F) \mod z^s \tag{7.10}$$

for any top level fiber component F in X, see (4.13). By (i) of Lemma 7.1.4, $\tilde{u} \equiv u_F \mod z^s$ in U_F. Due to condition (α_0) all the components of $f^*(0)$ in $\mathcal{X}$ are on the top level. Therefore, τ_* transforms the ideal $I = (v, f) \subset \mathcal{O}_{\mathcal{X}}(\mathcal{X})$ into $I' = (\tilde{u}, f) \subset \mathcal{O}_{\mathcal{X}}(\mathcal{X})$ preserving the principal ideal (f). By Lemma 1.4.3, τ induces an isomorphism $\tilde{\tau}: \mathcal{X}' \xrightarrow{\cong_B} \mathcal{X}''$ which gives (7.9), and so, proves our claim.

Let a be the maximal order of zeros of f. Letting $\tilde{u}' = \tilde{u}/f$ and $v'' = v/f$ one obtains

$$\tau_*: (z, \tilde{u}', v'') \mapsto (z, v'', -\tilde{u}') \mod z^{s-a}. \tag{7.11}$$

Let $\mathcal{F}'' = \tilde{\tau}(\mathcal{F}') \subset \mathcal{X}''$. Consider the standard affine charts $U_{F'} \times \mathbb{A}^1$ in $\mathcal{X}'$ and $U_{F''} \times \mathbb{A}^1$ in $\mathcal{X}''$ with local coordinates $(z, u_{F'}, v')$ and $(z, u_{F''}, v'')$, respectively, where

$$u_{F''} = u_F, \quad v' = v, \quad \text{and} \quad u_{F'} = u_F/f \equiv \tilde{u}' \mod z^{s-a}. \tag{7.12}$$

From (7.10) and (7.11) one can deduce

$$\tilde{\tau}: \mathcal{X}' \xrightarrow{\cong_B} \mathcal{X}'', \qquad \tilde{\tau}_*: (z, u_{F'}, v') \mapsto (z, v'', -u_{F''}) \mod z^{s-a}.$$

Then by (7.8) and (7.12) one gets

$$\beta^{-1} \circ \tilde{\tau}: \mathcal{X}' \xrightarrow{\cong_B} \mathcal{X}, \quad (z, u_{F'}, v') \mapsto (z, v, -u_F) \mod z^{s-a},$$

and so,

$$\tau^{-1} \circ \beta^{-1} \circ \tilde{\tau}: \mathcal{X}' \xrightarrow{\cong_B} \mathcal{X}, \quad (z, u_{F'}, v') \mapsto (z, u_F, v) \mod z^{s-a}.$$

Thus, the isomorphism $\varphi := \tilde{\tau}^{-1} \circ \beta \circ \tau: \mathcal{X} \xrightarrow{\cong_B} \mathcal{X}'$ verifies (7.7) with s replaced by $s - a$. □

EXAMPLE 7.1.15 (*Danielewski surfaces*). In the case of the Danielewski surfaces X_n, $n \in \mathbb{N}$ (see Examples 3.5.1 and 7.1.3), Proposition 7.1.14 and Corollary 7.1.19 below provide an alternative proof of the Danielewski–Fieseler Theorem ([**17**], [**26**]) which says that the cylinders $X_n \times \mathbb{A}^1$, $n \in \mathbb{N}$, are all isomorphic whereas the surfaces X_n and X_m are not if $n \neq m$.

REMARK 7.1.16. If, by chance, a GDF surface $\pi: X \to B$ verifies condition (α_0) with respect to some trivializing sequence (2.7) and some marking $z \in \mathcal{O}_B(B)$, then X admits a free $\mathbb{G}_a$-action along the fibers of π. In this case the conclusion of Proposition 7.1.14 can be derived by applying the Danielewski trick, see Section 1.3. See also Proposition 8.0.3 for concrete examples.

Next we provide an equivariant version of Proposition 7.1.14 with a similar proof. To avoid a repetition we omit certain details of the proof.

PROPOSITION 7.1.17. *Let $\pi: X \to B$ be a marked GDF μ_d-surface with a marking $z \in \mathcal{O}_B(B) \setminus \{0\}$ of weight 1 verifying condition (α) of Proposition 7.1.14. Then for any $k \in \mathbb{Z}$ and $l \in \mathbb{N}$ there exist a μ_d-equivariant principal top level A-stretching $\sigma: X' \to X$ with $A = \operatorname{div} z^{ld}$ and a μ_d-equivariant isomorphism of cylinders $\varphi: \mathcal{X}(k) \xrightarrow{\cong_{\mu_d,B}} \mathcal{X}'(k)$ verifying (7.7) for any component F of $z^*(0)$ in X.*

PROOF. By virtue of Lemma 7.1.12(b) one may suppose that all components of $z^*(0)$ in X are on the same top level m, that is, the graph divisor $\mathcal{D}(\pi)$ verifies condition (α_0), see the proof of Proposition 7.1.14. Let $\tilde{u} \in \mathcal{O}_X(X)$ be a μ_d-quasi-invariant of weight $-m$ satisfying conditions (i) and (ii) of Corollary 3.4.2 with $\mathfrak{F}_0$ being the set of (top level) components of $z^*(0)$ in X.

The iterated Asanuma modification of the second kind $\mathcal{X}''(k) \to \mathcal{X}(k)$, $v'' \mapsto v = z^{ld} v''$, is μ_d-equivariant along with the isomorphism

$$\beta_k: \mathcal{X}(k) \xrightarrow{\cong_{\mu_d,B}} \mathcal{X}''(k), \qquad v \mapsto v''.$$

By Lemma 4.4.3 one can find a μ_d-equivariant automorphism $\tau \in \operatorname{SAut}_B \mathcal{X}(-m)$ which interchanges modulo z^s, up to a sign, the functions $\tilde{u}$ and v (of the same weight $-m$) and leaves z invariant. By Lemma 1.4.4, τ admits a lift to a μ_d-equivariant isomorphism

$$\tilde{\tau}: \mathcal{X}'(-m) \xrightarrow{\cong_{\mu_d,B}} \mathcal{X}''(-m)$$

verifying (7.11) for $f = z^{ld}$. Now the composition $\varphi_{-m} = \tilde{\tau}^{-1} \circ \beta_{-m} \circ \tau$ yields an isomorphism $\mathcal{X}(-m) \xrightarrow{\cong_{\mu_d,B}} \mathcal{X}'(-m)$ verifying (7.7). By Lemma 5.4(c) one may replace the weight $-m$ by a given weight k. This does not affect (7.7) up to replacing s by $s-d$. □

We use below the following auxiliary fact.

LEMMA 7.1.18. *Let X be a normal affine surface that admits an $\mathbb{A}^1$-fibration $X \to C$ over a smooth affine curve C, and let $\bar{X} \to \bar{C}$ be a pseudominimal completion of $X \to C$ with extended graph Γ_{ext}. Then the number $v(\Gamma_{\text{ext}})$ of vertices of Γ_{ext} does not depend on the choice of an $\mathbb{A}^1$-fibration on X over an affine base. So, $v(X) := v(\Gamma_{\text{ext}})$ is an invariant of X.*

PROOF. Recall (see [**33**, Def. 2.16]) that every feather component F of the extended divisor D_{ext} is born under a blowup at a smooth point of the boundary divisor $D = \bar{X} \setminus X$. The unique component D_i of D containing the center of this blowup is called the *mother component* of F. The *normalization procedure* as defined in [**33**, Def. 3.2] replaces Γ_{ext} by the *normalized extended graph* $\Gamma_{\text{ext,norm}}$ such that any feather component F in Γ_{ext} becomes an extremal (-1)-vertex in $\Gamma_{\text{ext,norm}}$ attached at its mother component D_i. Under this procedure the total number of vertices remains the same: $v(\Gamma_{\text{ext,norm}}) = v(\Gamma_{\text{ext}})$. Furthermore, these graphs are assumed to be *standard*; the standardization procedure does not affect the number of vertices either, see [**33**, §1]. By [**33**, Thm. 3.5] the standard normalized extended graph $\Gamma_{\text{ext,norm}}$ of X is unique, that is, it does not depend on the choice of an $\mathbb{A}^1$-fibration on X over an affine base, unless X is a Gizatullin surface. In the latter case the minimal dual graph Γ of D is linear and $\Gamma_{\text{ext,norm}}$ is unique up to a *reversion* $\Gamma_{\mathrm{ext,norm}} \rightsquigarrow \Gamma^\vee_{\mathrm{ext,norm}}$. However, the reversion neither changes the number of vertices in Γ nor does it in $\Gamma_{\mathrm{ext,norm}}$. The latter is due to the *Matching Principle* ([**33**, Thm. 3.11]). According to this principle there is a bijection between the feather components of $\Gamma_{\mathrm{ext,norm}}$ and $\Gamma^\vee_{\mathrm{ext,norm}}$ along with their mother components. In conclusion, $v(\Gamma_{\mathrm{ext}}) = v(\Gamma_{\mathrm{ext,norm}})$ is an invariant of the surface X. □

COROLLARY 7.1.19. *Let $\theta : X' \to X$ be a principal geometric top-level A-stretching between two GDF surfaces $\pi': X' \to B$ and $\pi: X \to B$ where $A = \operatorname{div} f$, $f \in \mathcal{O}_X(X) \setminus \{0\}$. Suppose that $f(b) = 0$ for some $b \in B$ such that the fiber $\pi^{-1}(b)$ is reducible. Then $X' \not\cong X$.*

PROOF. Consider a trivializing completion $(\hat{X}, D)$ for $\pi: X \to B$ as in 2.4.7.1. Suppose that its degenerate fibers are situated over the points $b_1, \dots, b_n \in B$. This completion is pseudominimal if and only if for any $i \in \{1, \dots, n\}$ the only (-1)-vertices of $\Gamma_{b_i}(\pi)$ are leaves, that is, the root v_i of the fiber tree $\Gamma_{b_i}(\pi)$ is not a (-1)-vertex, see Lemma 2.3.8(c). If v_i is a (-1)-vertex then v_i is a tip of an extremal linear branch $\mathcal{B}_i$ of $\Gamma_{b_i}(\pi)$. There is an alternative: either $\Gamma_{b_i}(\pi) = \mathcal{B}_i$ is a chain $[[-1, -2, \dots, -2, -1]]$, or $\mathcal{B}_i$ is a branch $[[-1, -2, \dots, -2]]$ at a branching vertex v of $\Gamma_{b_i}(\pi)$ with weight ≤ -3. In the former case, $\Gamma_{b_i}(\pi)$ can be contracted to a single (0)-vertex. In the latter case, contracting $\mathcal{B}_i$ in $\Gamma_{b_i}(\pi)$ one gets a pseudominimal tree $\Gamma_{b_i,\min}(\pi)$ with v as the root. Performing these contractions of the branches $\mathcal{B}_i$ for all $i = 1, \dots, n$ yields a pseudominimal SNC completion $(\bar{X}_{\min}, D_{\min})$ of X.

Notice that

- $\mathcal{B}_i \subset \Gamma_{b_i}(\pi) \subset \Gamma_{b_i}(\pi')$;
- $\Gamma_{b_i}(\pi)$ is a chain if and only if the fiber $\pi^{-1}(b_i)$ is irreducible, if and only if $\Gamma_{b_i}(\pi')$ is.

By our assumption the fiber $\pi^{-1}(b_i)$ is reducible for some $i \in \{1, \dots, n\}$. Therefore, $\Gamma_{b_i}(\pi)$ is not a chain. Then the (eventual) contraction of $\mathcal{B}_i$ in both $\Gamma_{b_i}(\pi)$ and $\Gamma_{b_i}(\pi')$ leads to two pseudominimal rooted trees $\Gamma_{b_i,\min}(\pi)$ and $\Gamma_{b_i,\min}(\pi')$ where the number of vertices in $\Gamma_{b_i,\min}(\pi')$ is larger by $a_i = \operatorname{ord}_{b_i}(f) > 0$ than the one in $\Gamma_{b_i,\min}(\pi)$. In turn, the simultaneous contractions in all special fibers as described above lead to the pseudominimal completions $(\bar{X}_{\min}, D_{\min})$ and $(\bar{X}'_{\min}, D'_{\min})$ of $\pi: X \to B$ and $\pi': X' \to B$, respectively, such that the corresponding extended graphs $\Gamma_{\mathrm{ext,min}}$ and $\Gamma'_{\mathrm{ext,min}}$ have different number of vertices. By Lemma 7.1.18 this number is invariant upon isomorphisms of affine surfaces. Hence $X \not\cong X'$, as claimed. □

7.2. Non-cancellation for GDF surfaces

The main result of this section is the following theorem.

THEOREM 7.2.1. *Let $\pi: X \to B$ be a GDF surface. Then X is a Zariski 1-factor if and only if $\pi: X \to B$ admits a line bundle structure.*

PROOF. The 'if' part follows from Proposition 6.1.1. As for the 'only if' part, see the following version of Proposition 7.1.14 which avoids assumption (α). □

PROPOSITION 7.2.2. *Suppose that a GDF surface $\pi: X \to B$ has a reducible fiber. Then there exists a sequence of pairwise non-isomorphic GDF surfaces $\pi_{Y_k}: Y_k \to B$ with cylinders $\mathcal{Y}_k = Y_k \times \mathbb{A}^1$ isomorphic over B to $\mathcal{X} = X \times \mathbb{A}^1$: $\mathcal{Y}_k \cong_B \mathcal{X}$ $\forall k \in \mathbb{N}$.*

PROOF. Fix a well ordered trivializing sequence (2.7) with $X_m = X$. Let $l \in \{1,\dots,m\}$ be the minimal index such that $\pi_l: X_l \to B$ in (2.7) has a reducible fiber, say, $\pi_l^{-1}(b_1)$ where $b_1 \in B$. So, the graph divisor $\mathcal{D}(\pi_{l-1})$ is a chain divisor, and $\pi_l: X_l \to B$ verifies condition (α) of Proposition 7.1.14.

Consider a principal top-level A-stretching $\sigma: X_l' \to X_l$ where $A = \operatorname{div} z$. According to Proposition 7.1.14 for any $s \gg 1$ there exists an isomorphism of cylinders $\varphi_l: \mathcal{X}_l \xrightarrow{\cong_B} \mathcal{X}_l'$ satisfying (7.7). To any component $F \subset X_l$ of $\pi_l^{-1}(b_i)$ there corresponds a unique component $F' \subset X_l'$ of $\pi_l'^{-1}(b_i)$ such that $\sigma(F') = P_{F'} \in F$. Inspecting the proof of Proposition 7.1.14 we see that $\varphi_l(\mathcal{F}) = \mathcal{F}'$ where $\mathcal{F} = F \times \mathbb{A}^1 \subset \mathcal{X}_l$ and $\mathcal{F}' = F' \times \mathbb{A}^1 \subset \mathcal{X}_l'$. Due to (7.7) one has

$$\varphi_l(F \times \{0\}) = F' \times \{0\}. \tag{7.13}$$

Let us construct by recursion a sequence of GDF surfaces

$$X_m' \xrightarrow{\varrho_m'} X_{m-1}' \longrightarrow \dots \longrightarrow X_{l+1}' \xrightarrow{\varrho_{l+1}'} X_l' \tag{7.14}$$

such that

(i) $\mathcal{X}_j' \cong_B \mathcal{X}_j$, $j = l,\dots,m$;
(ii) $X_j' \not\cong X_j$ for any $j = l,\dots,m$.

Let $\Sigma = \bigcup_F \Sigma_F$ be the center of the affine modification $\varrho_{l+1}: X_{l+1} \to X_l$ from (2.7) where $\Sigma_F = \Sigma \cap F$. Set

$$\mathfrak{F}_0 = \{F \mid \Sigma_F \neq \varnothing\}, \qquad \mathfrak{F}_0' = \{F' \mid F \in \mathfrak{F}_0 \quad \text{and (7.13) holds}\},$$

$$\Sigma_{F'} \times \{0\} = \varphi_l(\Sigma_F \times \{0\}) \subset F' \times \{0\},$$

and

$$\Sigma' = \bigcup_{F' \in \mathfrak{F}_0'} \Sigma_{F'} \subset X_l'.$$

Consider the fibered modification $\varrho_{l+1}': X_{l+1}' \to X_l'$ along the reduced divisor $z^*(0)$ with the reduced center $\Sigma' \subset X_l'$. Let $\tilde\varrho_{l+1}: \mathcal{X}_{l+1} \to \mathcal{X}_l$ and $\tilde\varrho_{l+1}': \mathcal{X}_{l+1}' \to \mathcal{X}_l'$ be the Asanuma modifications of the first kind which correspond to ϱ_{l+1} and ϱ_{l+1}', respectively, see Lemma 5.1.1(a). By construction, $\varphi_l: \mathcal{X}_l \xrightarrow{\cong_B} \mathcal{X}_l'$ sends the center and the divisor of $\tilde\varrho_{l+1}$ to the center and the divisor of $\tilde\varrho_{l+1}'$. By Lemma 1.4.3, φ_l lifts to an isomorphism of cylinders $\varphi_{l+1}: \mathcal{X}_{l+1} \xrightarrow{\cong_B} \mathcal{X}_{l+1}'$. Likewise in (1.1) of Lemma 1.4.4, φ_{l+1} verifies (7.7) with s replaced by $s-1$. Now one can apply the same argument to the isomorphism $\varphi_{l+1}: \mathcal{X}_{l+1} \longrightarrow \mathcal{X}_{l+1}'$ instead of $\varphi_l: \mathcal{X}_l \longrightarrow \mathcal{X}_l'$. By recursion, we arrive at a sequence (7.14) such that $\mathcal{X}_i \cong_B \mathcal{X}_i'$ for all $i = 1,\dots,m$.

Let $Y_1 = X_m'$ and $\pi_{Y_1} = \pi_{X_m'}$. Then $\mathcal{Y}_1 \cong_B \mathcal{X}$. To finish the proof it suffices to repeat the same construction with z^k instead of z. This leads to a sequence of GDF surfaces $Y_k \to B$, $k = 1,2,\dots$. The number of vertices of the pseudominimal extended graphs $\Gamma^{(k)}_{\text{ext,min}}$ associated with Y_k strictly increases with k. Applying Lemma 7.1.18 in the same way as this is done in the proof of Corollary 7.1.19, one can see that the surfaces Y_k, $k = 1,2,\dots$, are pairwise non-isomorphic, while by construction their cylinders are all isomorphic to $\mathcal{X}$ over B. □

The following is an equivariant version of Proposition 7.2.2.

PROPOSITION 7.2.3. *Let $\pi: X \to B$ be a marked GDF μ_d-surface which has a reducible fiber. Then there exists a sequence of pairwise non-isomorphic marked*

GDF μ_d-surfaces $X^{(kd)}$, $k \in \mathbb{N}$, whose cylinders are μ_d-equivariantly isomorphic over B:

$$\mathcal{X}^{(kd)} \cong_{\mu_d,B} \mathcal{X} \quad \forall k \in \mathbb{N}\,.$$

PROOF. Consider a sequence (2.7) of μ_d-equivariant morphisms, and let $\pi_l: X_l \to B$ be the first member of (2.7) which has a reducible fiber. Proceeding as in the proof of Proposition 7.2.2 we let $X_l^{(kd)}$ be the GDF μ_d-surface obtained from X_l via a principal equivariant top-level A_k-stretching where $A_k = \operatorname{div} z^{kd}$, $k \in \mathbb{N}$.

According to Proposition 7.1.17 for any $s \gg 1$ there is a μ_d-equivariant isomorphism of cylinders $\varphi_l: \mathcal{X}_l(-l) \overset{\cong_{\mu_d,B}}{\longrightarrow} \mathcal{X}_l^{(kd)}(-l)$ satisfying (7.7). Repeating for any $k \in \mathbb{N}$ the construction from the proof of Proposition 7.2.2 in a μ_d-equivariant fashion one arrives at a sequence of marked GDF μ_d-surfaces $X^{(kd)} = X_m^{(kd)}$ with μ_d-equivariantly isomorphic cylinders $\mathcal{X}^{(kd)}(-m) \cong_{\mu_d,B} \mathcal{X}(-m)$ where the isomorphisms satisfy (7.7). By Lemma 5.1.4(c) there are μ_d-equivariant isomorphisms $\mathcal{X}^{(kd)}(0) \cong_{\mu_d,B} \mathcal{X}(0)$.

Arguing as in the proof of Corollary 7.1.19 one can see that the number of vertices of the corresponding pseudominimal extended graphs $\Gamma^{(kd)}_{\text{ext,min}}$ strictly increases with k. By Corollary 7.1.19 the GDF surfaces $X^{(kd)}$, $k = 0, 1, \ldots$, are pairwise non-isomorphic. □

7.3. Extended graphs of Gizatullin surfaces

The covering trick can be extended to completions as follows.

7.3.1 (*Covering trick for a completion*). Let $\pi_Y: Y \to C$ be an $\mathbb{A}^1$-fibration over an affine curve C, and let $\bar\pi_Y: \bar Y_{\text{resolved}} \to \bar C$ be a pseudominimal resolved completion of $\pi_Y: Y \to C$ with extended graph Γ_{ext}. Contracting the exceptional divisor $E \subset \bar Y_{\text{resolved}}$ of the minimal resolution of singularities of Y yields a birational morphism $\sigma: \bar Y_{\text{resolved}} \to \bar Y$ where $\bar Y \to \bar C$ is a completion of $\pi_Y: Y \to C$ with a simple normal crossing boundary divisor. Extending a branched covering $B \to C$ as in 2.1.2 to the smooth completions $\bar B \to \bar C$ consider the normalizations of the cross-products $\bar Y_{\text{resolved}} \times_{\bar C} \bar B$ and $\bar Y \times_{\bar C} \bar B$, the respective minimal desingularizations $\hat X_{\text{resolved}} \to (\bar Y_{\text{resolved}} \times_{\bar C} \bar B)_{\text{norm}}$ and $\hat X \to (\bar Y \times_{\bar C} \bar B)_{\text{norm}}$, and the induced $\mathbb{P}^1$-fibrations $\hat X_{\text{resolved}} \to \bar B$ and $\hat X \to \bar B$. Recall that the branched covering construction applied to $\pi_Y: Y \to C$ gives a GDF surface $\pi_X: X \to B$ as in Definition 2.1.2. The surface X is smooth, see Lemma 2.3.8(b). Hence $\hat X \to \bar B$ is a completion of $X \to B$ dominated by $\hat X_{\text{resolved}} \to \bar B$. The induced morphism $\hat\sigma: \hat X_{\text{resolved}} \to \hat X$ contracts the total transform of the exceptional divisor E of $\sigma: \bar Y_{\text{resolved}} \to \bar Y$.

7.3.2. Recall (see e.g., [**33**] and Section 1.1) that a Gizatullin surface X is a normal affine surface of class (ML_0). Such a surface X admits two different $\mathbb{A}^1$-fibrations over the affine line $\mathbb{A}^1$; in particular, $X \not\cong \mathbb{A}^1_* \times \mathbb{A}^1$. For any $\mathbb{A}^1$-fibration $\pi: X \to C$ over a smooth affine curve C one has $C \cong \mathbb{A}^1$ and π has at most one degenerate fiber. One may assume that this is the fiber $\pi^{-1}(0)$.

In the proof of Theorem 7.4.1 we use the following fact.

LEMMA 7.3.3. *Let X be a Gizatullin surface. Then the following hold.*

(a) *Let $\Omega(X)$ stand for the set of isomorphism classes of the pseudominimal extended graphs Γ_{ext} of all possible $\mathbb{A}^1$-fibrations $\pi: X \to \mathbb{A}^1$ with the special* *fiber* $\pi^{-1}(0)$. *Then* $\Omega(X)$ *is finite. Hence there exists* $d \in \mathbb{N}$ *such that the multiplicities of the fiber components of* $\pi^*(0)$ *in any such fibration divide* d.

(b) *Given an* $\mathbb{A}^1$*-fibration* $X \to \mathbb{A}^1 = C$ *consider the GDF surface* $\tilde{X} \to B$ *obtained via the cyclic base change* $\mathbb{A}^1 = B \to C$, $z \mapsto z^d$ *with* d *as in (a) and a subsequent normalization. Let* $\tilde{\Gamma}_{\text{ext}}$ *be the extended graph of a pseudominimal completion of* $\tilde{X}$. *Then the set* $\tilde{\Omega}(X,d)$ *of isomorphism classes of the graphs* $\tilde{\Gamma}_{\text{ext}}$ *for all possible* $\mathbb{A}^1$*-fibrations* $X \to \mathbb{A}^1 = C$ *is finite.*

PROOF. (a) By Lemma 7.1.18 all graphs Γ_{ext} in $\Omega(X)$ have the same number $v(X)$ of vertices. Notice that the number of non-isomorphic graphs on a given set of vertices is finite. Furthermore, given an $\mathbb{A}^1$-fibration $\pi: X \to \mathbb{A}^1$ along with a pseudominimal resolved completion $\bar{\pi}: \bar{X} \to \bar{\mathbb{P}}^1$ the multiplicities of the fiber components of $\bar{\pi}^{-1}(0)$ can be deduced in a combinatorial way from the associated extended graph Γ_{ext}. Hence there is $d \in \mathbb{N}$ divisible by all these multiplicities for all possible $\mathbb{A}^1$-fibrations $\pi: X \to \mathbb{A}^1$.

To show (b) it suffices to restrict to the $\mathbb{A}^1$-fibrations on X with a fixed pseudominimal extended graph Γ_{ext}. Let $\pi: X \to C = \mathbb{A}^1$ be such an $\mathbb{A}^1$-fibration. Recall that $B \cong C \cong \mathbb{A}^1$, the only possible degenerate fiber of π is $\pi^{-1}(0)$, and the base change $B \to C$ is $z \mapsto z^d$.

By 7.3.1 the extended graph $\hat{\Gamma}_{\text{ext}}$ is dominated by $\hat{\Gamma}_{\text{ext,resolved}}$. In turn, the pseudominimal extended graph $\tilde{\Gamma}_{\text{ext}}$ in $\tilde{\Omega}(X,d)$ is dominated by $\hat{\Gamma}_{\text{ext}}$ and also by $\hat{\Gamma}_{\text{ext,resolved}}$. This yields an upper bound for the number of vertices $v(\tilde{\Gamma}_{\text{ext}}) \le v(\hat{\Gamma}_{\text{ext,resolved}})$. We claim that $v(\hat{\Gamma}_{\text{ext,resolved}})$ is bounded above by a function depending only on d and on Γ_{ext}, and so, only on X, as desired.

To show the claim notice that for any vertex of Γ_{ext} there is at most d vertices of $\hat{\Gamma}_{\text{ext,resolved}}$ such that the corresponding curves in $\hat{X}_{\text{resolved}}$ dominate the one in $\bar{X}_{\text{resolved}}$. Hence it suffices to find an upper bound on the number of the remaining vertices of $\hat{\Gamma}_{\text{ext,resolved}}$ which correspond to the curves in $\hat{X}_{\text{resolved}}$ contracted in $\bar{X}_{\text{resolved}}$.

Let E' and E'' be two fiber components of the extended divisor D_{ext} with respective multiplicities m' and m'' that meet in $\bar{X}$. Choose local coordinates (x,y) in $\bar{X}$ centered at the intersection point $E' \cap E'' = \{p\}$ with E' and E'' as the axes. Then the germ of the cross-product $\bar{X} \times_{\bar{C}} \bar{B}$ near p is given by equation $z^d - x^{m'}y^{m''} = 0$. Likewise, if $E'' = S$ then the germ of $\bar{X} \times_{\bar{C}} \bar{B}$ near p is given by equation $z^d = x^{m'}$. Normalizing such a surface germ produces, in both cases, at most d cyclic quotient singularities of type uniquely determined by d, m', and m''. The resolution graphs of these singular points are the Hirzebruch-Jung strings uniquely determined by d and Γ_{ext}. It follows that the number of vertices in the total preimage in $\hat{\Gamma}_{\text{ext,resolved}}$ of the edge $[E, E']$ of Γ_{ext} is bounded above in terms of d and Γ_{ext}. Finally, $v(\hat{\Gamma}_{\text{ext,resolved}})$ is bounded above by a function of d and Γ_{ext}, as claimed. □

REMARK 7.3.4. It follows that a Gizatullin surface X admits at most a finite number of maximal families of pairwise non-equivalent $\mathbb{A}^1$-fibrations $X \to \mathbb{A}^1$.

Given $n \in \mathbb{N}$ there exists a finite number of maximal deformation families $\mathcal{F}_n \to \mathcal{S}_n$ of $\mathbb{A}^1$-fibrations $X_n \to \mathbb{A}^1$. The $\mathbb{A}^1$-fibrations in such a family are pairwise non-equivalent modulo the $(\operatorname{Aut} X_n)$-action, and $\dim \mathcal{S}_n$ strictly grows with n. All these families were described constructively in ([**32**, Cor. 6.3.20 and Ex. 6.3.21]. Taking in [**32**, Cor. 6.3.20] $n = m^2$, $k = m$, $a = 1$ and $l = m+3$ one obtains a family $\mathcal{F}_n \to \mathcal{S}_n$ such that the number of moduli $\dim \mathcal{S}_n$ has an optimal growth rate $n - 2\sqrt{n} - 1$.

7.4. Zariski 1-factors and affine $\mathbb{A}^1$-fibered surfaces

The following is the main result of Section 7.

THEOREM 7.4.1. *Let $\pi: X \to C$ be a normal $\mathbb{A}^1$-fibered affine surface over a smooth affine curve C. Then X is a Zariski 1-factor if and only if $\pi: X \to C$ admits a structure of a parabolic $\mathbb{G}_m$-surface.*

The "if" part follows from Theorem 6.3.1. The "only if" part is proven in the next proposition.

PROPOSITION 7.4.2. *Let $\pi: X \to C$ be an $\mathbb{A}^1$-fibration on a normal affine surface X over a smooth affine curve C. If X is a Zariski 1-factor then $\pi: X \to C$ admits a structure of a parabolic $\mathbb{G}_m$-surface.*

PROOF. Assume that $X \to C$ does not admit a structure of a parabolic $\mathbb{G}_m$-surface. We are going to construct an infinite sequence of normal affine surfaces $X^{(nd)}$ non-isomorphic to X such that the cylinders $\mathcal{X}^{(nd)}$ and $\mathcal{X}$ are isomorphic, thus showing that X cannot be a Zariski 1-factor.

Consider all possible $\mathbb{A}^1$-fibrations $X \to Z$ on X over smooth affine curves Z along with their pseudominimal extended graphs Γ_{ext}. By Lemma 7.3.3(a) the set $\Omega(X)$ of the isomorphism classes of graphs Γ_{ext} is finite. So, there is $d \in \mathbb{N}$ which divides the multiplicities of the fiber components in any $\mathbb{A}^1$-fibration $X \to Z$.

Applying to the given $\mathbb{A}^1$-fibration $X \to C$ the branched covering trick of Lemma 2.1.3 of degree d one obtains a marked GDF μ_d-surface $\tilde{X} \to B$. By Proposition 6.2.4, $\tilde{X} \to B$ does not admit a line bundle structure. By Proposition 7.2.3 there is a sequence of pairwise non-isomorphic marked GDF μ_d-surfaces $\tilde{X}^{(kd)} \to B$ such that for all $k \in \mathbb{N}$ the cylinders $\tilde{\mathcal{X}}^{(kd)}(0)$ and $\tilde{\mathcal{X}}(0)$ are μ_d-equivariantly isomorphic over B while $v(\tilde{X}^{(kd)}) \to \infty$ as $k \to \infty$ where v stands as before for the number of vertices in the extended graph of a pseudominimal completion. Passing to the quotients under the μ_d-actions yields a sequence of $\mathbb{A}^1$-fibered normal affine surfaces $X^{(kd)} = \tilde{X}^{(kd)}/\mu_d \to C$ with cylinders isomorphic over C:

$$\tilde{\mathcal{X}}^{(kd)}(0)/\mu_d \cong_C \tilde{\mathcal{X}}(0)/\mu_d = \mathcal{X} \quad \forall k \in \mathbb{N}.$$

We claim that under our assumptions the surface $X^{(kd)}$ is not isomorphic to X for any $k \gg 1$. Suppose to the contrary that $X^{(kd)} \cong X$ for an infinite set I of values of $k \geq 1$. Then X admits at least two different $\mathbb{A}^1$-fibrations over affine bases, that is, X is a Gizatullin surface. Indeed, otherwise any isomorphism $\varphi: X^{(kd)} \xrightarrow{\cong} X$ sends the $\mathbb{A}^1$-fibration $X^{(kd)} \to C$ to the unique $\mathbb{A}^1$-fibration $X \to C$. It can be lifted via the base change $B \to C$ and a normalization to an isomorphism $\tilde{\varphi}: \tilde{X}^{(kd)} \xrightarrow{\cong} \tilde{X}$. This gives a contradiction since $v(\tilde{X}^{(kd)}) > v(\tilde{X})$ for $k \gg 1$.

Thus, under our assumptions X and also $X^{(kd)} \cong X$, $k \in I$, are Gizatullin surfaces. Hence $C \cong \mathbb{A}^1$, and so, $B \cong \mathbb{A}^1$ too. By Lemma 7.3.3(b) the set $\tilde{\Omega}(X, d) = \tilde{\Omega}(X^{(kd)}, d)$, $k \in I$, is finite. In particular, for any $k \in I$ the pseudominimal extended graph $\tilde{\Gamma}^{(kd)}_{\text{ext}}$ associated with the GDF surface $\tilde{\pi}: \tilde{X}^{(kd)} \to B$ (which is a cyclic cover of $X^{(kd)}$) belongs to the finite set $\tilde{\Omega}(X^{(kd)}, d) = \tilde{\Omega}(X, d)$. Since the set $I \subset \mathbb{N}$ is infinite this contradicts the fact that $v(\tilde{\Gamma}_{\text{ext}}^{(kd)}) = v(\tilde{X}^{(kd)}) \to \infty$ as $k \to \infty$. Hence $X \not\cong X^{(kd)}$ for all $k \gg 1$. □

CHAPTER 8

Classical examples

In this chapter we analyze from our viewpoint examples of non-cancellation due to Danielewski [**17**], Fieseler [**26**], Wilkens [**69**], tom Dieck [**68**], and Miyanishi–Masuda [**54**]. We retrieve certain classification results for Danielewski-Fieseler surfaces in $\mathbb{A}^3$ due to Dubouloz and Poloni ([**25**], [**61**]) and derive new explicit examples of non-cancellation, see Proposition 8.0.3. See also [**27**, §4] and [**53**] for further examples.

NOTATION 8.0.1. Recall that a *bush* is a rooted tree such that all branches at the root vertex are chains, see Remark 6.2.5.2. Let $\Gamma_{d,m}$ be the bush with $d \geq 1$ branches of equal lengths $m \geq 1$. Let $\pi_{d,m}: X_{d,m} \to B = \operatorname{Spec} \Bbbk[z] \cong \mathbb{A}^1$ be a Danielewski-Fieseler surface with the unique special fiber $\pi_{d,m}^{-1}(0)$ such that $\Gamma_0(\pi_{d,m}) \cong \Gamma_{d,m}$. Thus, the fiber $\pi_{d,m}^*(0)$ is reduced and has d components $F_1, \ldots, F_d$, all on level m.

EXAMPLE 8.0.2. Let $X_{d,m,g}$ be a surface given in $\mathbb{A}^3 = \operatorname{Spec} \Bbbk[z, u, t]$ by an equation of the form

$$(8.1) \quad z^m t - g_m(z, u) = 0 \quad \text{where} \quad g_m(z, u) = b_0(u) + b_1(u) z + \ldots + b_{m-1}(u) z^{m-1}$$

for some $b_i \in \Bbbk[u]$, $i = 1, \ldots, m-1$ with $\deg b_i \leq d - 1$ and some monic polynomial $b_0 \in \Bbbk[u]$ of degree d with simple roots. Then $X_{d,m,g}$ is a surface of type $X_{d,m}$, see [**25**, §3.1]. The surfaces $X_{d,m,g}$ were classified in [**61**, Thm. 6.1(1)]; see also [**16**], [**52**], and [**58**] for some particular cases. For the reader's convenience we reproduce this classification with a new proof.

PROPOSITION 8.0.3. (a) *For any surface $X_{d,m}$ as in* 8.0.1 *there exists a polynomial $g = g_m \in \mathbb{K}[z, u]$ as in Example* 8.0.2 *such that $X_{d,m} \cong_B X_{d,m,g}$. Vice versa, any surface $X_{d,m,g}$ as in* 8.0.2 *is a Danielewski-Fieseler surface $X_{d,m}$ as in* 8.0.1.

(b) *Up to isomorphism over $B = \operatorname{Spec} \Bbbk[z]$ the cylinder $\mathcal{X}_{d,m,g} =: \mathcal{X}_{d,m}$ does not depend on the choice of $g = g_m$ in* (8.1). *Furthermore, $\mathcal{X}_{d,m} \cong_B \mathcal{X}_{d',m'}$ if and only if $d = d'$.*

(c) *If $X_{d,m,g} \cong X_{d',m',h}$ then $d = d'$ and $m = m'$. For $d, m \geq 2$ the following are equivalent:*
- *$X_{d,m,g} \cong X_{d,m,h}$;*
- *there is a commutative diagram*

$$(8.2) \qquad \begin{array}{ccc} X_{d,m,g} & \xrightarrow{\cong} & X_{d,m,h} \\ \downarrow & & \downarrow \\ B & \xrightarrow{\cong} & B \end{array}$$

- *there exist* $\alpha, \lambda \in \Bbbk^*$ *and* $\beta, \gamma \in \Bbbk[z]$ *with* $\deg \beta \le m-1$ *such that*
$$h(z,u) = (g(\lambda z, \alpha u + \beta(z)) - \gamma(z))/\alpha^d$$
where γ *is uniquely defined by* d, m, α, λ, *and* β *and the affine transformation* $u \mapsto \alpha u + \beta(0)$ *sends the roots of* $b_0 = g(0,u)$ *to the roots of* $c_0 := h(0,u)$.

PROOF. Consider a pseudominimal SNC completion $(\bar{X}_{d,m}, D_{d,m})$ of $X_{d,m}$ with projection $\bar{\pi}_{d,m}: \bar{X}_{d,m} \to \bar{B} = \mathbb{P}^1$ extending $\pi_{d,m}$, the fiber at infinity $\bar{F}_\infty \subset D_{d,m}$, the section at infinity $S \subset D_{d,m}$, and the unique reduced degenerate fiber $\bar{\pi}^*(0) = C_0 + \sum_{i=1}^d \mathcal{B}_i$ where C_0 is the root of $\Gamma_0(\pi_{d,m}) \cong \Gamma_{d,m}$ of weight $C_0^2 = -d$, and $\mathcal{B}_i = C_{i,1} + \ldots + C_{i,m-1} + \bar{F}_i$, $i = 1, \ldots, d$ are chains of length m with the sequence of weights $[[-2, \ldots, -2, -1]]$ and the (-1)-tips $\bar{F}_i$.

Let $\sigma: \bar{X}_{d,m} \to \bar{X}_{d,m-1}$ be the contraction of $\bar{F}_1, \ldots, \bar{F}_d$. The image of $C_{i,m-1}$ acquires the weight -1 on $\bar{X}_{d,m-1}$. Iterating this procedure leads to a sequence (2.8):

$$\bar{X}_{d,m} \xrightarrow{\bar{\varrho}_m} \bar{X}_{d,m-1} \longrightarrow \ldots \longrightarrow \bar{X}_{d,1} \xrightarrow{\bar{\varrho}_1} \bar{X}_{d,0} = \mathbb{P}^1 \times \mathbb{P}^1 \tag{8.3}$$

along with the corresponding trivializing sequence (2.7) for $X_{d,m}$ where $\bar{X}_{d,l}$ is a pseudominimal completion of $X_{d,l}$ and
$$\pi_{d,0}: X_{d,0} = \mathbb{A}^2 \to B = \mathbb{A}^1, \qquad (z,u) \mapsto z\,.$$
The affine modification $X_{d,1} \to X_{d,0} = \operatorname{Spec} \Bbbk[z,u]$ has divisor $z = 0$ and a reduced center consisting of d distinct points, say, $(0,\alpha_1), \ldots, (0,\alpha_d)$. Let $b_0 \in \Bbbk[u]$ be the monic polynomial of degree d with simple roots $\alpha_1, \ldots, \alpha_d$. One has
$$X_{d,1} \cong_B X_{d,1,b_0} \quad \text{where} \quad X_{d,1,b_0} = \{zt_1 - b_0(u) = 0\} \subset \mathbb{A}^3$$
is given by (8.1) with $m = 1$ and $t = t_1$.

Assume by recursion that $X_{d,l} \cong_B X_{d,l,g_l}$ where X_{d,l,g_l} is given in $\mathbb{A}^3 = \operatorname{Spec} \Bbbk[z,u,t_l]$ by an equation $z^l t_l - g_l(z,u) = 0$ as in (8.1). The surface $X_{d,l+1}$ in (2.7) is obtained from $X_{d,l} = X_{d,l,g_l}$ by a fibered modification with the reduced divisor $z^*(0)$ consisting of d disjoint components, say, $F_{1,l}, \ldots, F_{d,l} \subset X_{d,l,g_l}$ (where the closure $\bar{F}_{i,l}$ in $\bar{X}_{d,l}$ is the image of $C_{i,l}$) and a reduced center Z which consists of d points, say,
$$x_i = (0, \alpha_i, \beta_i) \in F_{i,l}, \quad i = 1, \ldots, d\,.$$
Let $b_l \in \Bbbk[u]$ be the polynomial of the minimal possible degree (so, $\deg b_l \le d-1$) such that $b_l(\alpha_i) + \beta_i = 0$, $i = 1, \ldots, d$. The three surfaces in $\mathbb{A}^3 = \operatorname{Spec} \Bbbk[z,u,t_l]$,
$$X_{d,l,g_l}, \quad \{z = 0\}, \quad \text{and} \quad \{b_l(u) + t_l = 0\}$$
meet transversely at the points x_i, $i = 1, \ldots, d$. Hence for the ideal $I \subset \mathcal{O}_{X_{d,l}}(X_{d,l})$ of Z one has $I = (z, b_l(u) + t_l)$. It follows that
$$\mathcal{O}_{X_{d,l+1}}(X_{d,l+1}) = \mathcal{O}_{X_{d,l}}(X_{d,l})[t_{l+1}] \quad \text{where} \quad t_{l+1} = (b_l(u) + t_l)/z\,,$$
see Definition 1.4.1. By the inductive hypothesis one obtains
$$t_l = g_l(z,u)/z^l = (b_0(u) + b_1(u)z + \ldots + b_{l-1}(u)z^{l-1})/z^l$$
where t_l fits in (8.1) with $m = l$. Therefore, $X_{d,l+1} = X_{d,l+1,g_{l+1}}$ is given in $\mathbb{A}^3 = \operatorname{Spec} \Bbbk[z,u,t_{l+1}]$ by the equation
$$z^{l+1} t_{l+1} - g_{l+1}(z,u) = 0 \quad \text{where} \quad g_{l+1}(z,u) = b_0(u) + b_1(u)z + \ldots + b_l(u)z^l\,.$$

This proves the first assertion in (a). Repeating the same argument in the reversed order gives the second (alternatively, see [**25**, §3.1]).

By (a) two GDF surfaces $X_{d,m,g}$ and $X_{d,m,h}$ with the same d and m have the same graph divisor. According to Theorem 5.2.2 the cylinders over these surfaces are isomorphic over $B = \operatorname{Spec} \Bbbk[z]$.

For any $m' > m$ the affine modification

$$\sigma_{m',m}: X_{d,m'} \to X_{d,m}, \qquad \sigma_{m',m} = \varrho_{m+1} \circ \cdots \circ \varrho_{m'}$$

is a principal top level stretching. Hence by Proposition 7.1.14 one has $\mathcal{X}_{d,m} \cong_B \mathcal{X}_{d,m'}$. Alternatively, the latter follows by the Danielewski's argument, see Remark 7.1.16.

The number d of components of the divisor $z^*(0)$ in $\mathcal{X}_{d,m}$ is invariant upon isomorphisms of cylinders over B. This yields (b).

The extended graph $\Gamma_{\text{ext}} = \bar{F}_\infty \cup S \cup \Gamma_{d,m}$ of the pseudominimal completion $(\bar{X}_{d,m}, D_{d,m})$ of $X_{d,m}$ has $v(X_{d,m}) = dm + 3$ vertices. According to Lemma 7.1.18, $v(X_{d,m})$ is an invariant of the surface, as well as the Picard number $\varrho(X_{d,m}) = d-1$, see (6.3). Hence $X_{d,m} \cong X_{d',m'}$ implies $dm = d'm'$ and $d = d'$, and so, $m = m'$.

For $d, m \geq 2$ the dual graph of the boundary divisor $\hat{D}_{d,m}$ is minimal and nonlinear. It follows that $X_{d,m}$ is a surface of class (ML_1), that is, it admits a unique $\mathbb{A}^1$-fibration over an affine base. Hence any isomorphism $X_{d,m,g} \cong X_{d,m,h}$ fits in a commutative diagram (8.2).

Next we follow the line of the proof of Lemma 5.12 in [**31**]. Assume there is an isomorphism $\varphi: X_{d,m,g} \xrightarrow{\cong} X_{d,m,h}$. The induced birational map $\bar{\varphi}: \bar{X}_{d,m,g} \dashrightarrow \bar{X}_{d,m,h}$ fits in the commutative diagram

$$\begin{array}{ccc} \bar{X}_{d,m,g} & \overset{\bar{\varphi}}{\dashrightarrow} & \bar{X}_{d,m,h} \\ \sigma \downarrow & & \downarrow \sigma' \\ W & \overset{\bar{\psi}}{\dashrightarrow} & W' \\ \downarrow & & \downarrow \\ \mathbb{P}^1 & \underset{\eta}{\overset{\cong}{\longrightarrow}} & \mathbb{P}^1 \end{array}$$

where σ is the contraction of the disjoint union of chains[1] $\mathcal{B}_i \ominus \bar{F}_i$ on $\bar{X}_{d,m,g}$, $i = 1, \ldots, d$ with a sequence of weights $[[-2, \ldots, -2]]$ of length $m-1$ along with the root C_0 to a normal surface singularity of W, and σ' is a similar contraction on $\bar{X}_{d,m,h}$. Clearly, $\bar{\varphi}$ is regular in the points of $S \setminus (C_0 \cup \bar{F}_\infty)$. So, $\bar{\psi}$ is biregular over $\mathbb{P}^1 \setminus \{\infty\}$ outside the isolated normal singularities of W and W'. By the Riemann extension theorem, $\bar{\psi}$ extends across the singular point of W yielding a biregular isomorphism $\bar{\psi}: W \setminus \sigma(\bar{F}_\infty) \to W' \setminus \sigma'(\bar{F}'_\infty)$. It follows that $\bar{\varphi}$ also extends to the minimal resolutions of singularities yielding an isomorphism $\bar{X}_{d,m,g} \setminus \bar{F}_\infty \xrightarrow{\cong} \bar{X}_{d,m,h} \setminus \bar{F}'_\infty$.

Contracting on the surfaces $\bar{X}_{d,m,g} \setminus (\bar{F}_\infty \cup S)$ and $\bar{X}_{d,m,h} \setminus (\bar{F}'_\infty \cup S')$ the unions of chains $\bigcup_{i=1}^d \mathcal{B}_i$ and $\bigcup_{i=1}^d \mathcal{B}'_i$, respectively, yields an isomorphism

$$\varphi: X_{d,0,g} = \operatorname{Spec} \Bbbk[z,u] \xrightarrow{\cong} X_{d,0,h} = \operatorname{Spec} \Bbbk[z,u], \quad (z,u) \mapsto (\lambda z, \alpha u + \beta(z))$$

[1]For a subgraph Γ' of a graph Γ we let $\Gamma \ominus \Gamma'$ denote the graph obtained from Γ by deleting Γ' along with the edges of Γ incident to Γ'.

where $\alpha, \lambda \in \Bbbk^*$ and $\beta \in \Bbbk[z]$. It sends the roots of $b_0(u) = g(0,u)$ into the roots of $c_0(u) = h(0,u)$. Hence

$$c_0(u) = \alpha^{-d} b_0 \circ \varphi(0,u) = \alpha^{-d} b_0(\alpha u + d_0) .$$

The automorphism $\varphi \in \operatorname{Aut} \mathbb{A}^2$ extends to an automorphism $\Phi \in \operatorname{Aut} \mathbb{A}^3$,

$$\Phi\colon (z,u,t) \mapsto \left(\lambda z, \alpha u + \beta(z), \frac{\alpha^d}{\lambda^m} t + \gamma(z)\right), \tag{8.4}$$

with a unique polynomial $\gamma \in \Bbbk[z]$ depending on $d, m, \alpha, \beta, \lambda$ such that the resulting equation $z^d t - h(z,u) = 0$ of the surface $X_{d,m,h} = \Phi(X_{d,m,g})$ has again the required form (8.1). The uniqueness of γ follows from the relation

$$\alpha^d h(z,u) = g(\lambda z, \alpha u + \beta(z)) - \lambda^m \gamma(z) .$$

This relation shows as well that $h(z,u)$ is independent of the higher order terms of $\beta(z)$. The rest of the proof of (c) is easy and can be left to the reader. □

8.0.4. Proposition 8.0.3 was established by Wilkens ([**69**]) for $\Bbbk = \mathbb{C}$, $d = 2$, $m \geq 2$, and $g(z,u)$ in (8.1) of the form $g(z,u) = h(z)u + u^2$ where $h \in \Bbbk[z]$ is a polynomial of degree $\deg h < m$ with $h(0) \neq 0$. Generalizing the examples of Danielewski-Fieseler [**17, 26**] and tom Dieck [**68**], Miyanishi and Masuda ([**54**]) considered yet another instance of surfaces in $\mathbb{A}^3_{\mathbb{C}}$ given by (8.1).

EXAMPLE 8.0.5 (*Masuda-Miyanishi* [**54**]). Given natural d, m with $d > m \geq 2$ and a homogeneous polynomial

$$g(z,u) = u^d + a_2 u^{d-2} z^2 + \ldots + a_d z^d \tag{8.5}$$

consider the surface $X(d,m,g)$ in $\mathbb{A}^3$ defined by the equation

$$z^m t - g(z,u) - 1 = 0 .$$

Then $\pi = z|_{X(d,m,g)}\colon X(d,m,g) \to B = \operatorname{Spec} \Bbbk[z]$ is a Danielewski-Fieseler surface with a unique special fiber $\pi^{-1}(0)$ consisting of d components on level m. The examples of tom Dieck ([**68**]) correspond to the case $g(z,u) = u^d$. By a suitable triangular automorphism $(z,u,t) \mapsto (z,u,t-b(z,u))$ one can eliminate the last $d-m$ terms in (8.5) and reduce the equation to (8.1). The next result of Miyanishi and Masuda is a particular case of Proposition 8.0.3.

PROPOSITION 8.0.6 ([**54**, Thm. 2.8]). *With the notation as in* 8.0.5 *the following hold.*

(a) $\mathcal{X}(d,m,g) \cong_B \mathcal{X}(d,m',h)$ *for any* d, m, m' *and* $g, h \in \Bbbk[z,u]$ *as in* (8.5).
(b) $X(d,m,g) \cong X(d,m',h)$ *if and only if* $m = m'$ *and* $h(z,u) = g(\lambda z, u)$ *for some* $\lambda \in \Bbbk^*$.

CHAPTER 9

GDF surfaces with isomorphic cylinders

This chapter is devoted to the proof of Theorem 0.0.6 in the Introduction; see also Theorem 9.2.1 below. Hereafter B stands for a smooth affine curve.

9.1. Preliminaries

9.1.1. Recall (see Definition 7.1.8) that the Danielewski-Fieseler quotient $\mathrm{DF}(\pi)$ of a GDF surface $\pi: X \to B$ is a one-dimensional scheme obtained as the quotient of X by the equivalence relation defined by the fiber components of π. The type divisor $\mathrm{tp.div}(\pi)$ is an anti-effective divisor on $\mathrm{DF}(\pi)$ taking the value $-l(F)$ for a special fiber component F of π on level $l(F)$, see Definition 7.1.9.

Remind also that the linear equivalence of divisors on $\mathrm{DF}(\pi)$ is, as usual, the equivalence modulo the principal divisors on $\mathrm{DF}(\pi)$. The latter divisors are just the principal divisors on B lifted to $\mathrm{DF}(\pi)$. The Picard group $\mathrm{Pic}\,\mathrm{DF}(\pi)$ is defined in a usual way. To a GDF surface $\pi: X \to B$ we associate its Picard class $[\mathrm{tp.div}(\pi)] \in \mathrm{Pic}\,\mathrm{DF}(\pi)$. If $\pi: X \to B$ represents a line bundle L over B then $\mathrm{DF}(\pi) = B$ and $[\mathrm{tp.div}(\pi)] = [L] \in \mathrm{Pic}\,B$.

REMARKS 9.1.2. 1. Let $q: \mathrm{DF}(\pi) \to B$ be the natural projection, and let $b_1, \ldots, b_n \in B$ be the points such that $N_j := \mathrm{card}\, q^{-1}(b_j) > 1$. It is easily seen that

$$\mathrm{Pic}\,\mathrm{DF}(\pi) \cong (\mathrm{Pic}\,B) \oplus \mathbb{Z}^{\varrho(\pi)} \quad \text{where} \quad \varrho(\pi) = \sum_{j=1}^{n} (N_j - 1).$$

2. Let $\pi: X \to B$ be a GDF surface. It is well known that the natural projection $p: X \to \mathrm{DF}(\pi)$ admits a structure of an $\mathbb{A}^1$-fiber bundle (see, e.g., Proposition 3.2.2). If $\mathrm{DF}(\pi)$ is non-separated then the latter fiber bundle does not admit any regular section. Indeed, the image of such a section would be a non-separated reduced proper subscheme of X. However, the latter is not possible since X is separated. By a similar reason, given a line bundle L over $\mathrm{DF}(\pi)$, the total space of L is affine if and only if $\mathrm{DF}(\pi)$ is separated, that is, $\mathrm{DF}(\pi) = B$.

3. Let as before F be a special fiber component of π on level $l(F)$. Let $b_F = p(F) \in \mathrm{DF}(\pi)$, $b_i = \pi(F) = q(b_F) \in B$, and $B_i = B \setminus \{b_1, \ldots, b_n\} \cup \{b_i\}$.

Given a regular section $s_0: B \to X_0 = B \times \mathbb{A}^1$ in (2.7) the proper transform s_m of s_0 in $X = X_m$ acquires a pole of order $l(F)$ over b_F. The latter means the following. Choose the standard affine chart $U_F \cong_{B_i} B_i \times \mathbb{A}^1$ about F with natural local coordinates (z, u_F) where $z \in \mathcal{O}_B(B)$ is a marking giving a local parameter on (B_i, b_i). Then at the point b_i the rational function $z^{l_F} u_F \circ s_m$ is regular and does not vanish. Therefore, $-\mathrm{div}_\infty(s_m) \sim \mathrm{tp.div}(\pi)$ where the pole divisor $\mathrm{div}_\infty(s_m)$ on $\mathrm{DF}(\pi)$ is defined above. The proof goes by induction on the length l_F of the shortest path in the fiber tree $\Gamma_{b_i}(\pi)$ joining the vertex $\bar{F}$ with the root. Each subsequent blowup in (2.7) along this path increases the pole order by one.

EXAMPLE 9.1.3. If all fibers of a GDF surface $\pi: X \to B$ are irreducible then π is the projection of a locally trivial $\mathbb{A}^1$-fiber bundle. Since the base B is affine this fiber bundle admits regular sections. Given such a section Z there is a unique structure of a line bundle, say, L on $\pi: X \to B$ with zero section Z. Let $Z' \neq Z$ be a second regular section of L, and let L' be the line bundle over B with projection π and zero section Z'. Since the effective zero divisor $\operatorname{div}_Z(Z') = \operatorname{div}_{Z'}(Z)$ belongs to the classes of both L and of L' in $\operatorname{Pic} B$, these classes coincide, that is, $L \cong_B L'$.

Notice that any class in $\operatorname{Pic} B$ contains effective and anti-effective divisors. Via the above procedure any such class can be represented by an $\mathbb{A}^1$-fiber bundle $X \to B$, that is, by a GDF surface with only irreducible fibers.

9.2. Classification of GDF cylinders up to B-isomorphism

Theorem 0.0.6 can be reformulated as follows.

THEOREM 9.2.1. *Let $\pi_X: X \to B$ and $\pi_Y: Y \to B$ be two GDF surfaces over the same base B. Then the cylinders $\mathcal{X} = X \times \mathbb{A}^1$ and $\mathcal{Y} = Y \times \mathbb{A}^1$ are isomorphic over B if and only if there exists an isomorphism $\tau: \mathrm{DF}(\pi_X) \xrightarrow{\cong_B} \mathrm{DF}(\pi_Y)$ such that $[\mathrm{tp.div}(\pi_X)] = [\tau^*(\mathrm{tp.div}(\pi_Y))]$ in the Picard group $\operatorname{Pic}\mathrm{DF}(\pi_X)$.*

The proof starts with the following elementary lemma.

LEMMA 9.2.2. *An isomorphism $\varphi: \mathcal{X} \xrightarrow{\cong_B} \mathcal{Y}$ induces an isomorphism*

$$\tau: \mathrm{DF}(\pi_X) \xrightarrow{\cong_B} \mathrm{DF}(\pi_Y).$$

PROOF. Recall (see Definition 7.1.8) that the Danielewski-Fieseler quotient $\mathrm{DF}(\pi_X)$ is the quotient of X by the equivalence relation defined by the fiber components of π_X. It coincides with the quotient of the cylinder $\mathcal{X}$ by the equivalence relation defined by the fiber components of $\mathcal{X} \to B$. The isomorphism $\varphi: \mathcal{X} \xrightarrow{\cong_B} \mathcal{Y}$ respects the latter equivalence relations on $\mathcal{X}$ and $\mathcal{Y}$. □

This lemma along with the linear equivalence of the type divisors established in Section 9.6.1 gives the 'only if' part of Theorem 9.2.1.

The strategy of the proof of the 'if' part is as follows. We reduce the assertion to the case where the type divisor completely determines the graph divisor. This is so indeed if the fiber trees are bushes, see Definition 9.3.1. First we obtain this reduction assuming that the GDF surface $\pi_X: X \to B$ has a unique special fiber $\pi_X^{-1}(b_0)$, $b_0 \in B$, see Theorem 9.5.3. In Section 9.6.2 we treat the general case of GDF surfaces with any number of special fibers.

Theorem 5.2.2 is crucial for the proof. This theorem says that certain continuous parameters of a GDF surface $\pi_X: X \to B$ are irrelevant for the B-isomorphism class of the cylinder $\mathcal{X} = X \times \mathbb{A}^1$. This allows to replace the initial GDF surface $\pi_X: X \to B$ by a suitable new one $\pi_{X'}: X' \to B$ which carries the same graph divisor $\mathcal{D}(\pi_{X'}) = \mathcal{D}(\pi_X)$ and admits a rather simple explicit description.

Let us indicate some corollaries of Theorem 9.2.1. First of all, the GDF surfaces over a given smooth affine curve B which are not Zariski 1-factors and whose cylinders are isomorphic over B to a given cylinder form a countable number of families with affine bases. The dimensions of the bases are unbounded. This follows also from Theorem 5.2.2, Proposition 7.1.14, Lemma 7.1.18, and Corollary 7.1.19. A similar conclusion holds as well for the collection of all normal affine surfaces $\mathbb{A}^1$-fibered over a given smooth affine curve and having a given cylinder.

The following result confirms Conjecture 1.2.1 under a certain additional assumption. It is due to Bandman and Makar-Limanov ([**9**, Thm. 1]).

THEOREM 9.2.3 (Bandman-Makar-Limanov). *Let* $\pi_X: X \to \mathbb{A}^1$ *be a Danielewski-Fieseler surface, that is, a GDF surface over* $B = \operatorname{Spec} \Bbbk[z]$ *with the unique special fiber* $z^*(0)$. *Then there is an isomorphism of cylinders* $\mathcal{X} \cong_B \mathcal{Y}$ *where* $\mathcal{Y} = Y \times \mathbb{A}^1$ *is the cylinder over a Gizatullin GDF surface* $\pi_Y: Y \to \mathbb{A}^1$. *Hence the group* $\operatorname{SAut} \mathcal{X}$ *acts in* $\mathcal{X}$ *with a Zariski open orbit* O *such that* $\operatorname{codim}_{\mathcal{X}}(\mathcal{X} \setminus O) \geq 2$.

An elegant direct proof of this theorem in [**9**] exploits the Danielewski construction. We provide here an alternative argument based on Theorem 9.2.1.

PROOF. The assertion is trivial if $X \cong \mathbb{A}^2$. To exclude this case we will assume that the fiber $\pi_X^{-1}(0)$ with the fiber tree $\Gamma_0 = \Gamma_0(\pi_X)$ is reducible. Letting $\operatorname{tp}(\Gamma_0) = (n_i)_{i=0,\dots,h}$, see 2.3.9, by our assumption $n_0 = 0$ and $h = \operatorname{ht}(\Gamma_0) > 0$.

By Theorem 9.2.1 it suffices to find a Gizatullin GDF surface $\pi_Y: Y \to \mathbb{A}^1$ with a unique special fiber $\pi_Y^{-1}(0)$ such that $\operatorname{tp}(\Gamma_0(\pi_Y)) = \operatorname{tp}(\Gamma_0)$. Consider a chain $\gamma_0 = [v_0, \dots, v_{h-1}]$ of length $h-1$. Let Γ_0' be the rooted tree with v_0 for the root, γ_0 for the trunk, and with the leaves $v_{i+1,j}$, $j = 1, \dots, n_{i+1}$ on level $i+1$, $i = 0, \dots, h-1$, joint with γ_0 by the edges $[v_i, v_{i+1,j}]$. So, Γ_0' has exactly n_i leaves $v_{i,j}$ on level i. Therefore, $\operatorname{tp}(\Gamma_0') = \operatorname{tp}(\Gamma_0)$.

Let $\pi_Y: Y \to \mathbb{A}^1$ be a Danielewski-Fieseler surface such that $\Gamma_0(\pi_Y) = \Gamma_0'$. By Theorem 9.2.1 one has $\mathcal{X} \cong_B \mathcal{Y}$. Clearly, $Y \not\cong \mathbb{A}^1 \times \mathbb{A}^1_*$ is a Gizatullin surface. Indeed, there exists an SNC completion $(\bar{Y}, D)$ of Y with the linear dual graph $\Gamma(D) = [F_\infty, S, v_0, \dots, v_{h-1}]$ obtained by attaching to γ_0 the $[[0,0]]$-chain $[F_\infty, S]$. Thus, Y is a Gizatullin surface.

The group $\operatorname{SAut} Y$ acts on Y with a Zariski open orbit whose complement is finite ([**39**]). Hence also $\operatorname{SAut} \mathcal{Y} \supset \operatorname{SAut} Y \times \operatorname{SAut} \mathbb{A}^1$ has a Zariski open orbit in $\mathcal{Y} \cong \mathcal{X}$ with a complement of codimension at least 2. In particular, $\operatorname{ML}(\mathcal{X}) \cong \operatorname{ML}(\mathcal{Y})$ is trivial. □

9.3. GDF surfaces whose fiber trees are bushes

In this section we assume for simplicity that B is the affine line $\mathbb{A}^1$ and $\pi: X \to \mathbb{A}^1$ has a unique special fiber $\pi^{-1}(0)$. Later on we will indicate the modifications which allow to treat the general case.

DEFINITION 9.3.1 (*Bushes*). Recall that a *bush* is a rooted tree Γ such that the branches of Γ at the root v_0 are chains, see Figure 9.1. If $\operatorname{tp}(\Gamma) = (n_i)_{i\geq 0}$, then Γ has exactly n_i branches of height i, the root v_0 of Γ being the common tip of each branch.

DEFINITION 9.3.2 (*Accompanying sequence of a bush*). Let Γ be a bush of height $m > 0$. We associate with Γ a sequence $(p_i, r_i)_{i=1,\dots,m} \in (\Bbbk[u])^{2m}$ of pairs of monic polynomials with simple roots such that

- $\deg p_i$ equals the number of vertices of Γ on level i;
- $p_{i+1} | p_i$ for $i = 1, \dots, m-1$;
- $r_i p_i = p_1$ for $i = 1, \dots, m$.

Thus, $r_1 = 1$ and $r_i | r_{i+1}$ for $i = 1, \dots, m-1$. Furthermore, $\operatorname{tp}(\Gamma) = (n_i)_{i\geq 0}$ where $n_i = \deg p_i - \deg p_{i+1}$; see Definition 2.3.9. For instance, for Γ in Figure 9.1 one can

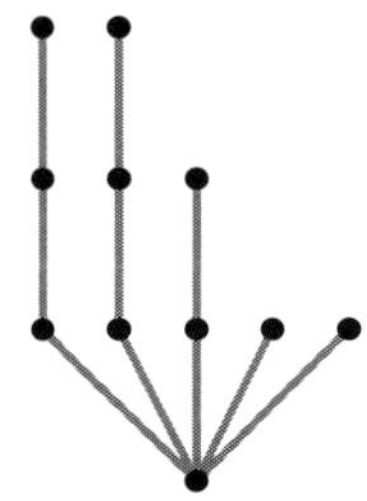

FIGURE 9.1. A bush Γ of height 3 and of type $\mathrm{tp}(\Gamma) = (0,2,1,2)$

choose

$$(9.1)\quad p_1 = \prod_{i=0}^{4}(u-i), \quad p_2 = \prod_{i=0}^{2}(u-i), \quad p_3 = \prod_{i=0}^{1}(u-i), \quad \text{and} \quad r_i = p_1/p_i,\ i = 1,2,3.$$

To any root α of p_1 there corresponds a unique branch $\mathcal{B}_\alpha$ of Γ of height $l(\alpha) := \mathrm{ht}(\mathcal{B}_\alpha) \geq 1$ where $l(\alpha)$ satisfies

$$(9.2)\qquad p_1(\alpha) = p_2(\alpha) = \ldots = p_{l(\alpha)}(\alpha) = 0 \quad \text{and} \quad p_{l(\alpha)+1}(\alpha) \neq 0$$

or, which is equivalent,

$$(9.3)\qquad r_1(\alpha) \neq 0, \ldots, r_{l(\alpha)}(\alpha) \neq 0, \quad r_{l(\alpha)+1}(\alpha) = \ldots = r_m(\alpha) = 0\,.$$

EXAMPLE 9.3.3. Consider the surface $X_m = \{z^m t - p(u) = 0\}$ in $\mathbb{A}^3$ of Danielewski type with the projection $\pi_m\colon X_m \to \mathbb{A}^1$, $(z,u,t) \mapsto z$. Then $\Gamma_0(\pi_m)$ is a bush with $d = \deg p$ branches of equal length m and $(p_i, r_i) = (p, 1)$ for $i = 1, \ldots, m$; cf. Examples 3.5.1, 3.5.2, and 7.1.15.

REMARK 9.3.4. The accompanying sequence $(p_i, r_i)_{i=1,\ldots,m}$ of Γ is uniquely determined by the pair (Γ, p_1). Given Γ this family of polynomials is parameterized by the coefficients of p_1, that is, by the points in $\mathbb{A}^n \setminus D_n$ where $n = \deg p_1$ and $D_n = \{\mathrm{discr}(p_1) = 0\}$ is the discriminant hypersurface in $\mathbb{A}^n$.

Hereafter we adopt the following convention.

CONVENTION 9.3.5 (*Reduction of the base field*). Given a finite collection of affine varieties over $\Bbbk$ defined by systems of polynomial equations in $\mathbb{A}^N$ one can replace the base field $\Bbbk$ by the finite extension $\mathbb{Q} \subset \Bbbk'$ generated in $\Bbbk$ by the coefficients of all these polynomials. Choosing an embedding $\Bbbk' \hookrightarrow \mathbb{C}$ one may assume that $\Bbbk = \mathbb{C}$.

In the next proposition we associate to any bush Γ_0 along with an accompanying sequence (p_i, r_i) a certain Danielewski-Fieseler surface $\pi\colon X \to \mathbb{A}^1$ in $\mathbb{A}^N$ such that $\Gamma_0(\pi) = \Gamma_0$; cf. [**19, 20, 25**].

PROPOSITION 9.3.6. *Fix a bush Γ_0 of height $m > 0$ and an accompanying sequence $(p_i, r_i)_{i=1,\ldots,m}$ as in 9.3.2. For $1 \leq j \leq m$ consider the subvariety $W_j \subset \mathbb{A}^{j+2} = \operatorname{Spec} \Bbbk[z, u, t_1, \ldots, t_j]$ given by*

$$(9.4)\qquad zt_1 - p_1(u) = 0, \quad zt_i - r_i(u)t_{i-1} = 0, \quad i = 2, \ldots, j\,.$$

Then the following hold.

(i) *There is a unique irreducible component X_j of W_j which dominates the z-axis;*

(ii) $\pi_j := z|_{X_j}: X_j \to \mathbb{A}^1$ *is a GDF surface with the unique special fiber* $\pi_j^{-1}(0)$;

(iii) $(X_j)_{j=1,\dots,m}$ *fits in a sequence* (2.7) *of fibered modifications*

$$X_m \xrightarrow{\varrho_m} X_{m-1} \longrightarrow \dots \longrightarrow X_1 \xrightarrow{\varrho_1} X_0 = \mathbb{A}^2 = \operatorname{Spec} \Bbbk[z,u]$$

where

$$\varrho_i: (z,u,t_1,\dots,t_i) \mapsto (z,u,t_1,\dots,t_{i-1}), \quad i = 1,\dots,m;$$

(iv) $\Gamma_0(\pi_j)$ *is the bush* Γ_0 *truncated at the level* j. *In particular,* $\Gamma_0(\pi_m) = \Gamma_0$.

PROOF. Consider the hyperplane $L_0 = \{z = 0\}$ in $\mathbb{A}^{j+2}$. The projection

$$\pi: \mathbb{A}^{j+2} \to \mathbb{A}^2, \quad (z,u,t_1,\dots,t_j) \mapsto (z,u) \tag{9.5}$$

restricted to $W_j \smallsetminus L_0$ yields an isomorphism $W_j \smallsetminus L_0 \cong \mathbb{A}^1_* \times \mathbb{A}^1$. The closure $X_j := \overline{W_j \smallsetminus L_0}$ is a surface satisfying (i) such that $\pi_j|_{X_j \smallsetminus L_0}: X_j \smallsetminus L_0 \to \mathbb{A}^1_*$ is induced by the first projection of $\mathbb{A}^1_* \times \mathbb{A}^1$. Any component of W_j different from X_j is of the form $\{(0,\alpha)\} \times \mathbb{A}^j$ where α is a root of p_1 and $\mathbb{A}^j = \operatorname{Spec} \Bbbk[t_1,\dots,t_j]$.

Let us introduce the following notation. Given a root α of p_1 consider the open set $\omega(\alpha) \subset \mathbb{A}^1 = \operatorname{Spec} \Bbbk[u]$ given by $p_1(u)/(u-\alpha) \neq 0$. Let $U_0(\alpha) = \mathbb{A}^1 \times \omega(\alpha) \subset X_0 = \operatorname{Spec} \Bbbk[z,u]$, and for $j = 1,\dots,m$ let $U_j(\alpha) = \pi^{-1}(U_0(\alpha)) \subset X_j$ where π is as in (9.5). The collection $\{U_j(\alpha)\}_{p_1(\alpha)=0}$ gives an open covering of X_j. Restricting to $U_j(\alpha)$ amounts to passing to the localization

$$\mathcal{O}_{U_j(\alpha)}(U_j(\alpha)) = \mathcal{O}_{X_j}(X_j)[(u-\beta)^{-1} | p_1(\beta) = 0, \beta \neq \alpha].$$

The projection $\pi_j|_{U_j(\alpha)}: U_j(\alpha) \to \mathbb{A}^1 = \operatorname{Spec} \Bbbk[z]$ is dominant and defines an $\mathbb{A}^1$-fibered surface such that $U_j(\alpha) \smallsetminus L_0 = X_j \smallsetminus L_0$.

CLAIM. Let (9.3) holds for a root α of p_1 with $l(\alpha) = l$. Then the pair $(z, t_j - t_j(x))$ for $j \in \{1\dots,l\}$ and $(z, t_l - t_l(x))$ for $j \in \{l,\dots,m\}$, respectively, is a local analytic coordinate system near any point $x \in X_j$ such that $z(x) = 0$ and $u(x) = \alpha$. In particular, (X_j, x) is a smooth surface germ. The divisor $\pi_j^*(0)$ is reduced, the fiber $\pi_j^{-1}(0)$ is smooth, and each of its components is isomorphic to $\mathbb{A}^1$.

PROOF OF THE CLAIM. Replacing u by $u - \alpha$ one may assume that $\alpha = 0$. all assertions but the last one are local, hence one may restrict to the neighborhood $U_j := U_j(0)$ of x.

We proceed by induction on j. Since $p_1(u)/u \in \Bbbk[u]$ does not vanish in U_0 one has by (9.4):

$$\mathcal{O}_{U_1}(U_1) = \mathcal{O}_{U_0}(U_0)[p_1(u)/z] = \mathcal{O}_{U_0}(U_0)[u/z].$$

Thus, $U_1 \to U_0$ is an affine modification of $U_0 \subset \mathbb{A}^2 = \operatorname{Spec} \Bbbk[z,u]$ along the reduced divisor $z^*(0)$ whose center is the ideal (z,u). This yields an embedding $U_1 \hookrightarrow \mathbb{A}^2 = \operatorname{Spec} \Bbbk[z,t_1']$ where $t_1' = u/z$ commuting with the projections to $\mathbb{A}^1 = \operatorname{Spec} \Bbbk[z]$ and sending the exceptional divisor to the axis $\{z = 0\}$ in $\mathbb{A}^2$. Since $t_1 = p_1(u)/z$ and t_1' differ in U_1 by the invertible factor $p_1(u)/u$ our assertions hold for $j = 1$.

Assume by recursion that for some $j < l$ there is an embedding $U_j \hookrightarrow \mathbb{A}^2 = \operatorname{Spec} \Bbbk[z,t_j']$ which commutes with the projections to $\mathbb{A}^1 = \operatorname{Spec} \Bbbk[z]$. Consider the map $U_{j+1} \to U_j$ forgetting the last coordinate $t_{j+1} = r_{j+1}(u)t_j/z$, see (9.4). Since $j + 1 \le l$ the function r_{j+1} is invertible in U_j, see (9.3). Hence

$$\mathcal{O}_{U_{j+1}}(U_{j+1}) = \mathcal{O}_{U_j}(U_j)[r_{j+1}(u)t_j/z] = \mathcal{O}_{U_j}(U_j)[t_j/z]. \tag{9.6}$$

This yields an embedding $U_{j+1} \hookrightarrow \mathbb{A}^2 = \operatorname{Spec} \Bbbk[z, t'_{j+1}]$ commuting with the projections to $\mathbb{A}^1 = \operatorname{Spec} \Bbbk[z]$ where $t'_{j+1} := t_j/z$ differs in U_{j+1} from t_{j+1} by an invertible factor r_{j+1}. Hence (z, t_{j+1}) gives local analytic coordinates near the axis $z = 0$. This proves the claim for $j = 1, \dots, l$.

The first equation in (9.4) gives $u/z = ut_1/p_1(u) \in \mathcal{O}_{U_j}(U_j)$. Hence $r_{j+1}(u)/z \in \mathcal{O}_{U_j}(U_j)$, and so, $\mathcal{O}_{U_{j+1}}(U_{j+1}) = \mathcal{O}_{U_j}(U_j)$ for all $j \in \{l, \dots, m-1\}$, see (9.6). This means that $U_{j+1} \to U_j$ is an isomorphism commuting with the projections to $\mathbb{A}^1 = \operatorname{Spec} \Bbbk[z]$. Now the claim follows.

According to the claim for any $j = 1, \dots, m$ the surface X_j is smooth, the fiber $\pi_j^*(0)$ is reduced, and the unique component $F_{\alpha,j} \subset X_j$ of this fiber with $u|_{F_{\alpha,j}} = \alpha$ is an $\mathbb{A}^1$-curve parameterized by t_l where $l = \min\{j, l(\alpha)\}$. This yields (ii).

It follows also that the morphism $\varrho_{j+1}\colon X_{j+1} \to X_j$ as in (iii) sends the affine line $F_{\alpha,j+1} \subset X_{j+1}$ isomorphically onto the affine line $F_{\alpha,j} \subset X_j$ for any $j = l, \dots, m-1$ and contracts $F_{\alpha,j+1}$ to the point $(0, \alpha, 0, \dots, 0) \in \mathbb{A}^{j+2}$ for any $j = 0, \dots, l-1$. This proves (iii) and shows as well that $F_{\alpha,j}$ is on level j for $1 \le j \le l-1$ and on level l for $l \le j \le m$. Hence one has $\Gamma_0(\pi_j) = (\Gamma_0)_{\le j}$ as stated in (iv). □

REMARK 9.3.7. It follows from the proof that $\Gamma_0(\pi_j|_{U_j(\alpha)}) = (\mathcal{B}_\alpha)_{\le j}$ for $j = 1, \dots, l(\alpha)$ and $\Gamma_0(\pi_j|_{U_j(\alpha)}) = \mathcal{B}_\alpha$ for $j = l(\alpha), \dots, m$. In particular, the graph $\Gamma_0(\pi_m|_{U_m(\alpha)})$ coincides with the branch $\mathcal{B}_\alpha$ of Γ_0 of height $\operatorname{ht}(\mathcal{B}_\alpha) = l(\alpha)$.

From Proposition 9.3.6 and its proof we deduce such a corollary.

COROLLARY 9.3.8. *We keep the notation as before. Given a root α of p_1 consider the fiber component $F_\alpha = F_{\alpha,m} = \{z = 0, u - \alpha = 0\} \subset \pi_m^{-1}(0)$ on X_m. Then the following hold.*

(1) *$t_{l(\alpha)}$ coincides with $\frac{u-\alpha}{z^{l(\alpha)}}$ up to a factor which is a regular and invertible function in a Zariski open neighborhood $U_m(\alpha) \subset X_m$ of F_α;*
(2) *$(z, t_{l(\alpha)})$ yields a local analytic coordinate system in X_m near F_α;*
(3) *for $i \in \{1, \dots, m\}$ the restriction $r_i(u)t_i|_{F_\alpha}$ vanishes if $i \ne l(\alpha)$ and gives an affine coordinate on $F_\alpha \cong \mathbb{A}^1$ if $i = l(\alpha)$.*

PROOF. Statements (a) and (b) follow by an argument in the proof of Proposition 9.3.6 (see the Claim).

(c) By (9.3) one has $r_i(\alpha) = 0$ for $i > l(\alpha)$ and $r_i(\alpha) \ne 0$ for $i \le l(\alpha)$. Due to (b), $r_i(\alpha)t_i|_{F_\alpha}$ is an affine coordinate if $i = l(\alpha)$. If $i < l(\alpha)$ then $r_{i+1}(\alpha) \ne 0$ and, by (9.4), $t_i|_{F_\alpha} = z\frac{t_{i+1}}{r_{i+1}(\alpha)}|_{F_\alpha} = 0$. □

9.4. Spring bushes versus bushes

In this section we extend the results of Section 9.3 to the Danielewski-Fieseler surfaces whose fiber trees are *spring bushes*.

DEFINITION 9.4.1 (*Spring bush*). A rooted tree $\widehat{\Gamma}$ of height $\hat{h} \ge 1$ is called a *spring bush* if the truncation $\Gamma := \widehat{\Gamma}_{\le \hat{h}-1}$ is a bush sharing the same root with Γ. Thus, any leaf of $\widehat{\Gamma}$ is a neighbor of a leaf of Γ. A spring bush is not necessarily a bush (see Figure 9.2).

DEFINITION 9.4.2 (*Accompanying sequence of a spring bush*). Let $\hat{\Gamma}$ be a spring bush of height $\hat{h} = \operatorname{ht}(\hat{\Gamma})$ over a bush Γ of height $h = \hat{h}-1$, and let $(p_i, r_i)_{i=1,\dots,h}$ be the

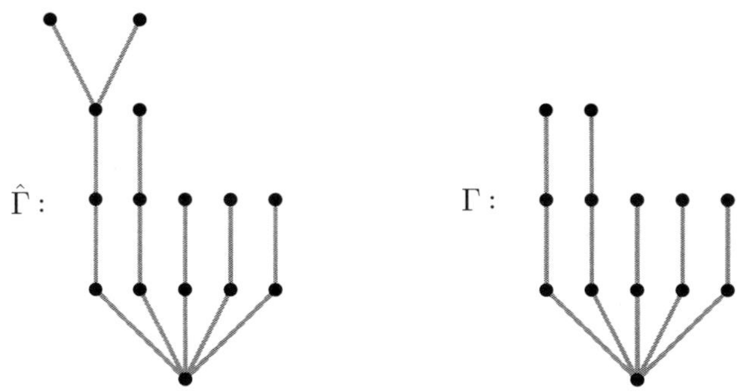

FIGURE 9.2. A spring bush $\hat{\Gamma}$ of type $\mathrm{tp}(\hat{\Gamma}) = (0, 0, 3, 1, 2)$ over a bush $\Gamma = \hat{\Gamma}_{\le 3}$

accompanying sequence of Γ. Recall that for each root α of p_1 there corresponds a unique linear branch $\mathcal{B}_\alpha$ of Γ (see Definition 9.3.2). Let $p_{\hat{h}}$ be the monic polynomial dividing p_h such that the roots of $p_{\hat{h}}$ correspond to the branches of Γ linked to the top vertices of $\hat{\Gamma}$. Let also $r_{\hat{h}} = p_1/p_{\hat{h}}$. For instance, for $\hat{\Gamma}$ in Figure 9.2 along with the accompanying sequence (9.1) of Γ and the root $\alpha_1 = 0$ of p_1 one has $(p_4, r_4) = (u, p_1/u)$.

For a branch $\mathcal{B}_\alpha$ of Γ linked to $n(\alpha)$ leaves of $\hat{\Gamma}$ we fix a monic polynomial with simple roots $q_\alpha \in \Bbbk[v]$ of degree $n(\alpha)$ where $q_\alpha(v) = v$ if $n(\alpha) = 1$. We let

$$q(u, v) = \sum_{p_1(\alpha)=0} q_\alpha(v) \frac{p_1(u)}{u - \alpha} \in \Bbbk[u, v]. \tag{9.7}$$

The system of polynomials $\{(p_i, r_i)_{i=1,\dots,\hat{h}}, q\}$ is called an *accompanying sequence of* $\hat{\Gamma}$.

Our main result in this section is the following.

PROPOSITION 9.4.3. *Let $\hat{\Gamma}$ be a spring bush, and let $\tilde{\Gamma}$ be a bush with $\mathrm{tp}(\widehat{\Gamma}) = \mathrm{tp}(\widetilde{\Gamma})$, see Figure 9.3. Letting $B = \mathrm{Spec}\, \Bbbk[z] \cong \mathbb{A}^1$ consider Danielewski-Fieseler surfaces $\pi_{\hat{X}}: \hat{X} \to B$ and $\pi_Y: Y \to B$ with the unique special fibers over $0 \in B$ such that $\Gamma_0(\hat{X}) = \hat{\Gamma}$ and $\Gamma_0(Y) = \tilde{\Gamma}$. Then there is an isomorphism of cylinders $\varphi: \hat{\mathcal{X}} \xrightarrow{\cong_B} \mathcal{Y}$ which sends the components of $z^*(0)$ in $\hat{\mathcal{X}}$ to components of $z^*(0)$ in $\mathcal{Y}$ preserving the levels.*

FIGURE 9.3. A spring bush $\hat{\Gamma}$ and a bush $\tilde{\Gamma}$ of the same type $(0, 1, 2)$

The proof is done at the end of this section. We start with the following elementary fact that can be interpreted in terms of commuting affine modifications, cf. Remark 1.4.2.2.

LEMMA 9.4.4. *Let A be an affine domain, and let $a_1,\dots,a_k,b_1,\dots,b_l \in \mathrm{Frac}(A)$ be elements of the field of fractions of A. Consider the extensions $A' = A[a_1,\dots,a_k]$, $B = A[b_1,\dots,b_l]$, and $B' = A[a_1,\dots,a_k,b_1,\dots,b_l]$. Then $B' = B[a_1,\dots,a_k]$, that is, the extensions $A \subset A'$ and $B \subset B'$ share the same system of generators $a_1,\dots,a_k$.*

REMARK 9.4.5. Let $\pi: X \to B$ be a marked GDF surface with a marking $z \in \mathcal{O}_B(B)$ where $z^*(0) = b_1 + \ldots + b_n$ is a reduced effective divisor. Performing a principal $(A,\overline{-1})$-stretching with the divisor $A = \operatorname{div} z^d = d(b_1 + \ldots + b_n)$ one may suppose that the graph divisor $\mathcal{D}(\pi)$ does not have any leaf on level $\le d-1$; cf. Lemma 7.1.11.

NOTATION 9.4.6. In the remaining part of this section we consider the following objects.

- $\hat{\Gamma}$ - a spring bush of height $\mathrm{ht}(\hat{\Gamma}) = m+1 \ge 3$ over a bush Γ along with an accompanying sequence $\{(p_i,r_i)_{i=1,\dots,m+1},q\}$ as in 9.4.2. According to Remark 9.4.5 we may and we will assume that Γ has no branch of height 1;
- Γ_h - the subbush of height $m+1$ of $\hat{\Gamma}$ obtained by removing on each branch of $\hat{\Gamma}$ of height $m+1$ all leaves but one, see Figure 9.4;
- Γ_s - the subbush of height m of Γ_h obtained from Γ_h by removing all the leaves of Γ_h, see Figure 9.4;
- $\mathcal{B}_s(\alpha)$ - the branch of Γ_s of height $l_s(\alpha) = \mathrm{ht}(\mathcal{B}_s(\alpha))$ which corresponds to a root α of p_1; $\mathcal{B}_h(\alpha) \subset \Gamma_h$ and $l_h(\alpha)$ are defined likewise;
- $B = \operatorname{Spec} \Bbbk[z] \cong \mathbb{A}^1$;
- $\hat{\pi}: \hat{X} \to B$, $\pi_h: X_h \to B$, and $\pi_s: X_s \to B$ - the Danielewski-Fieseler surfaces with a unique special fiber over $0 \in B$ such that $\Gamma_0(\hat{\pi}) = \hat{\Gamma}$, $\Gamma_0(\pi_h) = \Gamma_h$, and $\Gamma_0(\pi_s) = \Gamma_s$ where
- $X_h \subset \mathbb{A}^{m+3}$ and $X_s \subset \mathbb{A}^{m+2}$ satisfy the corresponding systems (9.4). So, (9.4) provides the extensions
$$\Bbbk[z,u] \subset \mathcal{O}_{X_s}(X_s) \subset \mathcal{O}_{X_h}(X_h);$$
- $F_s(\alpha) \subset X_s$ - the fiber component of $z^*(0)$ corresponding to the leaf of $\mathcal{B}_s(\alpha)$;
- $t_s(\alpha) = (u-\alpha)/z^{l_s(\alpha)}$ - an affine coordinate on $F_s(\alpha)$, see Corollary 9.3.8(a);
- $F_h(\alpha) \subset X_h$ and $t_h(\alpha)$ are defined likewise;
- $\hat{\mathcal{X}}, \mathcal{X}_h, \mathcal{X}_s$ - the cylinders over $\hat{X}, X_h$, and X_s, respectively;
- $\mathcal{F}_s(\alpha) = F_s(\alpha) \times \mathbb{A}^1 \cong \operatorname{Spec} \Bbbk[t_s(\alpha),v] \cong \mathbb{A}^2$ - the fiber component of $z^*(0)$ in $\mathcal{X}_s$ which corresponds to α; $\mathcal{F}_h(\alpha) \subset \mathcal{X}_h$ is defined likewise.

LEMMA 9.4.7. *There is an isomorphism $\hat{\mathcal{X}} \cong_B V$ where the affine threefold V results from the affine modification $V \to \mathcal{X}_h$ along divisor $z^*(0)$ on $\mathcal{X}_h$ with center the ideal $(z,q(u,v)) \subset \mathcal{O}_{\mathcal{X}_h}(\mathcal{X}_h)$ for $q \in \Bbbk[u,v]$ as in (9.7).*

PROOF. Let $t_s \in \mathcal{O}_{X_s}(X_s)$ be such that $t_s|_{F_s(\alpha)} = t_s(\alpha)$ for any root α of p_1. The extension

$$\mathcal{O}_{X_s}(X_s) \subset \mathcal{O}_{X_h}(X_h) = \mathcal{O}_{X_s}(X_s)[t_s/z] \tag{9.8}$$

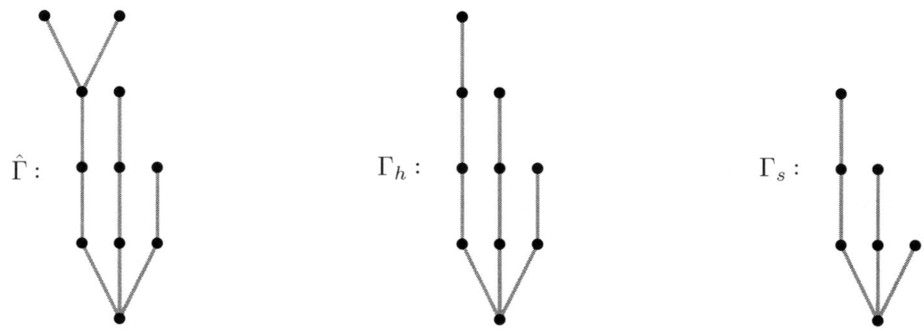

FIGURE 9.4. A spring bush $\hat{\Gamma}$ and its subbushes Γ_h and Γ_s

amounts to the fiber modification $\sigma_h: X_h \to X_s$ along divisor $z^*(0)$ whose center is the ideal $(z, t_s) \subset \mathcal{O}_{X_s}(X_s)$. The extension

$$\mathcal{O}_{X_s}(X_s) \subset \mathcal{O}_{\hat{X}}(\hat{X}) = \mathcal{O}_{X_s}(X_s)[q(u,t_s)/z] \tag{9.9}$$

amounts to the fiber modification $\hat{X} \to X_s$ along divisor $z^*(0)$ whose center is the ideal $(z, q(u,t_s)) \subset \mathcal{O}_{X_s}(X_s)$.

Letting $\mathcal{O}_{\mathcal{X}_s}(\mathcal{X}_s) = \mathcal{O}_{X_s}(X_s)[v]$, by Lemma 5.1.1 applied to (9.9) the cylinder $\hat{\mathcal{X}}$ can be obtained from $\mathcal{X}_s$ via the affine modification $\hat{\sigma}: \hat{\mathcal{X}} \to \mathcal{X}_s$ along divisor $z^*(0)$ whose center is the ideal $J_s = (z, q(u,t_s), v) \subset \mathcal{O}_{\mathcal{X}_s}(\mathcal{X}_s)$, that is,

$$\mathcal{O}_{\hat{\mathcal{X}}}(\hat{\mathcal{X}}) \cong_{\Bbbk[z]} \mathcal{O}_{\mathcal{X}_s}(\mathcal{X}_s)[q(u,t_s)/z, v/z]\,. \tag{9.10}$$

The ideal J_s defines a reduced zero dimensional subscheme $\mathcal{S} \subset \mathcal{X}_s$. Let

$$\mathcal{S}(\alpha) = \mathcal{S} \cap \mathcal{F}_s(\alpha) = \mathcal{F}_s(\alpha) \cap \{q_\alpha(t_s) = 0\}\,.$$

One has $\operatorname{card}\mathcal{S}(\alpha) = n(\alpha) \ge 1$, see (9.7). In the case $l_s(\alpha) < m$ one has $n(\alpha) = 1$ and $\mathcal{S}_\alpha = \{t_s = 0, v = 0\} \subset \mathcal{F}_s(\alpha)$. Let

$$\mathfrak{S}(\alpha) = \{t_s = 0, q_\alpha(v) = 0\} \subset \mathcal{F}_s(\alpha)\,.$$

One has $\operatorname{card}\mathcal{S}(\alpha) = \operatorname{card}\mathfrak{S}(\alpha)$ and, moreover, $\mathcal{S}(\alpha) = \mathfrak{S}(\alpha)$ if $l_s(\alpha) < m$, see 9.4.2. According to Theorem 4.1.4 the relative flexibility holds for the cylinder $\mathcal{X}_s$ (see Definition 4.1.2). Hence there is an automorphism $\tau \in \mathrm{SAut}_B(\mathcal{X}_s)$ such that

- $\tau(\mathcal{F}_s(\alpha)) = \mathcal{F}_s(\alpha)\ \forall \alpha$;
- $\tau|_{\mathcal{F}_s(\alpha)} = \mathrm{id}$ if $l_s(\alpha) < m$;
- $\tau(\mathcal{S}(\alpha)) = \mathfrak{S}(\alpha)$ if $l_s(\alpha) = m$.

Thus, $\tau(\mathcal{S}(\alpha)) = \mathfrak{S}(\alpha)$ for any α, and so, τ_* sends J_s to the ideal

$$I_s := (z, t_s, q(u,v)) \subset \mathcal{O}_{\mathcal{X}_s}(\mathcal{X}_s).$$

Consider further the affine threefold $W = \operatorname{Spec}\mathcal{O}_{\mathcal{X}_s}(\mathcal{X}_s)[t_s/z, q(u,v)/z]$. It results from the affine modification of $\mathcal{X}_s$ along divisor $z^*(0)$ whose center is the ideal I_s. By virtue of (9.10) and Lemma 1.4.3, τ admits a lift to an isomorphism $\hat{\mathcal{X}} \xrightarrow{\cong_B} W$. We claim that there is an isomorphism $W \cong_B V$. Indeed, using (9.8) and Lemma 9.4.4 one obtains

$$\begin{aligned}\mathcal{O}_W(W) &= \mathcal{O}_{\mathcal{X}_s}(\mathcal{X}_s)[t_s/z, q(u,v)/z] \\ &= \mathcal{O}_{\mathcal{X}_s}(\mathcal{X}_s)[t_s/z][q(u,v)/z] \cong_{\Bbbk[z]} \mathcal{O}_{\mathcal{X}_h}(\mathcal{X}_h)[q(u,v)/z] = \mathcal{O}_V(V)\,.\end{aligned}$$

□

In the proof of Proposition 9.4.3 we use the following Lemmas 9.4.8–9.4.11.

LEMMA 9.4.8. *We adopt Notation 9.4.6. To a trivializing sequence (2.7) for X_h:*

$$X_h = X_{h,m+1} \stackrel{\varrho_{m+1}}{\longrightarrow} X_{h,m} \longrightarrow \ldots \longrightarrow X_{h,1} \stackrel{\varrho_1}{\longrightarrow} X_{h,0} = B \times \mathbb{A}^1 \qquad (9.11)$$

there corresponds a trivializing sequence (5.3) of cylinders:

$$\mathcal{X}_h = \mathcal{X}_{h,m+1} \stackrel{\tilde\rho_{m+1}}{\longrightarrow} \mathcal{X}_{h,m} \longrightarrow \ldots \longrightarrow \mathcal{X}_{h,1} \stackrel{\tilde\rho_1}{\longrightarrow} \mathcal{X}_{h,0} = B \times \mathbb{A}^2 \qquad (9.12)$$

where $\tilde\rho_{i+1} = \varrho_{i+1} \times \mathrm{id}_{\mathbb{A}^1}$, $i = 0, \ldots, m$. Given a root α of p_1 let $F_{h,i}(\alpha)$ be the corresponding component of the divisor $z^*(0)$ on $X_{h,i}$ equipped with the affine coordinate

$$t_{h,i}(\alpha) = (u - \alpha) \cdot z^{-l_{h,i}(\alpha)}$$

where $(u-\alpha)|_{F_{h,i}(\alpha)} = 0$ and $l_{h,i}(\alpha)$ is the level of $F_{h,i}(\alpha)$, see Corollary 9.3.8(a). Then the affine modification $\tilde\rho_{i+1}\colon \mathcal{X}_{h,i+1} \to \mathcal{X}_{h,i}$ amounts to the extension

$$\mathcal{O}_{\mathcal{X}_{h,i}}(\mathcal{X}_{h,i}) \subset \mathcal{O}_{\mathcal{X}_{h,i+1}}(\mathcal{X}_{h,i+1}) = \mathcal{O}_{\mathcal{X}_{h,i}}(\mathcal{X}_{h,i})[t_{h,i}/z] \qquad (9.13)$$

where $t_{h,i} \in \mathcal{O}_{X_{h,i}}(X_{h,i})$ satisfies

$$t_{h,i}|_{F_{h,i}(\alpha)} = \begin{cases} t_{h,i}(\alpha), & l_h(\alpha) \ge i+1, \\ 0, & l_h(\alpha) \le i\,. \end{cases}$$

PROOF. The center of the affine modification $\tilde\rho_{i+1}\colon \mathcal{X}_{h,i+1} \to \mathcal{X}_{h,i}$ along divisor $z^*(0)$ on $\mathcal{X}_{h,i}$ is the union of the affine lines

$$L_{h,i}(\alpha) = \{t_{h,i}(\alpha) = 0\} \cong \operatorname{Spec} \Bbbk[v] \cong \mathbb{A}^1 \qquad (9.14)$$

such that $l_h(\alpha) \ge i+1$. Each component

$$\mathcal{F}_{h,i}(\alpha) = F_{h,i} \times \mathbb{A}^1 \cong \operatorname{Spec} \Bbbk[t_{h,i}(\alpha), v] \cong \mathbb{A}^2 \qquad (9.15)$$

of $z^*(0)$ in $\mathcal{X}_{h,i}$ contains at most one such line. Now the lemma follows easily. □

LEMMA 9.4.9. *Let $\mathrm{tp}(\hat\Gamma) = (n_0, \ldots, n_{m+1})$. Consider the polynomials*

$$q_\alpha \in \Bbbk[u], \quad \deg q_\alpha = n(\alpha), \quad \textit{and} \quad q \in \Bbbk[u,v]$$

as in (9.7), see 9.4.2. For each $i = 1, \ldots, m+1$ consider the affine threefold

$$V_i = \operatorname{Spec} \mathcal{O}_{\mathcal{X}_{h,i}}(\mathcal{X}_{h,i})[q(u,v)/z]$$

given in $\mathcal{X}_{h,i} \times \mathbb{A}^1$ by equation $zw - q(u,v) = 0$ where $\mathbb{A}^1 = \operatorname{Spec} \Bbbk[w]$. Then the natural projection $V_i \to \mathcal{X}_{h,i}$ is the affine modification along divisor $z^*(0)$ on $\mathcal{X}_{h,i}$ with the reduced center $\mathcal{L}_i = \cup_{\alpha,\beta} L_{\alpha,\beta}$ where the affine line

$$L_{\alpha,\beta} = \{v - \beta = 0\} \subset \mathcal{F}_{h,i}(\alpha) \cong \mathbb{A}^2,$$

see (9.15), corresponds to a solution $(\alpha,\beta) \in \mathbb{A}^2 = \operatorname{Spec} \Bbbk[u,v]$ of the system

$$p_1(u) = 0, \quad q(u,v) = 0\,.$$

The plane $\mathcal{F}_{h,i}(\alpha)$ contains $n(\alpha)$ such lines $L_{\alpha,\beta}$. The center of the modification $V_i \to \mathcal{X}_{h,i}$ consists of

$$N = \sum_{p_1(\alpha)=0} n(\alpha) = \sum_{k=0}^{m+1} n_k \qquad (9.16)$$

affine lines. The reduced divisor $z^*(0)$ *on* V_i *has* N *disjoint components* $\mathcal{F}_{i,(\alpha,\beta)}$ *where*

$$\mathcal{F}_{i,(\alpha,\beta)} = \{z = 0,\ u = \alpha,\ v = \beta\} \cong \operatorname{Spec} \Bbbk[t_{h,i}(\alpha), w] \cong \mathbb{A}^2 . \tag{9.17}$$

PROOF. The proof is straightforward. □

LEMMA 9.4.10. *There is a sequence*

$$V_{m+1} \xrightarrow{\nu_{m+1}} V_m \longrightarrow \ldots \longrightarrow V_2 \xrightarrow{\nu_2} V_1 \tag{9.18}$$

where $\nu_{i+1}: V_{i+1} \to V_i$ *is the affine modification along divisor* $z^*(0)$ *with a reduced center* $\mathfrak{S}_i$ *defined by the ideal* $(z, t_{h,i}) \subset \mathcal{O}_{V_i}(V_i)$ *and consisting of the*

$$N_i = \sum_{l_h(\alpha) \geq i+1} n(\alpha) = \sum_{k=i+1}^{m+1} n_k \tag{9.19}$$

affine lines $L_{i,(\alpha,\beta)}$ *with* $l_h(\alpha) \geq i + 1$.

PROOF. Using Lemma 9.4.4 and (9.13) one obtains

$$\begin{aligned}\mathcal{O}_{V_{i+1}}(V_{i+1}) &= \mathcal{O}_{\mathcal{X}_{h,i+1}}(\mathcal{X}_{h,i+1})[q(u,v)/z] = \mathcal{O}_{\mathcal{X}_{h,i}}(\mathcal{X}_{h,i})[t_{h,i}/z][q(u,v)/z] \\ &= \mathcal{O}_{\mathcal{X}_{h,i}}(\mathcal{X}_{h,i})[q(u,v)/z][t_{h,i}/z] = \mathcal{O}_{V_i}(V_i)[t_{h,i}/z] .\end{aligned}$$

□

LEMMA 9.4.11. *Let* $\pi_Y : Y \to B = \mathbb{A}^1$ *be a Danielewski-Fieseler surface as in Proposition 9.4.3, that is,* $\Gamma_0(Y)$ *is the bush* $\tilde{\Gamma}$ *with* $\operatorname{tp}(\tilde{\Gamma}) = \operatorname{tp}(\hat{\Gamma}) = (n_0, n_1, \ldots, n_{m+1})$. *Given a the trivializing sequence*

$$Y = Y_{m+1} \xrightarrow{\sigma_{m+1}} Y_m \longrightarrow \ldots \longrightarrow Y_1 \xrightarrow{\sigma_1} Y_0 = B \times \mathbb{A}^1 = \operatorname{Spec} \Bbbk[z, u] \tag{9.20}$$

with $\Gamma_0(\pi_{Y_i}) = \tilde{\Gamma}_{\leq i}$ *there is a trivializing sequence of cylinders*

$$\mathcal{Y} = \mathcal{Y}_{m+1} \xrightarrow{s_{m+1}} \mathcal{Y}_m \longrightarrow \ldots \longrightarrow \mathcal{Y}_1 \xrightarrow{s_1} \mathcal{Y}_0 = B \times \mathbb{A}^2 \tag{9.21}$$

where for $i \geq 1$ *the morphism* $: \mathcal{Y}_{i+1} \to \mathcal{Y}_i$, $s_{i+1} = \sigma_{i+1} \times \mathrm{id}_{\mathbb{A}^1}$ *is an affine modification along the reduced divisor* $z^*(0)$ *on* $\mathcal{Y}_i$ *with a reduced center* $\mathcal{S}_i$ *such that*

- $z^{-1}(0)$ *is a disjoint union of* N *components* $\tilde{\mathcal{F}}_{i,j} \cong \mathbb{A}^2$ *with* N *as in (9.16) where* $\tilde{\mathcal{F}}_{i,j}$ *corresponds to the vertex on level* $l(\tilde{\mathcal{F}}_{i,j})$ *of the branch* $\tilde{\mathcal{B}}_j$ *of* $\tilde{\Gamma}$;
- *the center* $\mathcal{S}_i$ *is a disjoint union of* N_i *affine lines* $\tilde{L}_{i,j} \subset \tilde{\mathcal{F}}_{i,j}$ *with* $\operatorname{ht}(\tilde{\mathcal{B}}_j) \geq i + 1$ *where* N_i *is as in (9.19) and* $\mathcal{S}_i \cap \mathcal{F}_{i,j}$ *consists of at most one line* $L_{i,j}$.

PROOF. Since $\operatorname{tp}(\tilde{\Gamma}) = \operatorname{tp}(\hat{\Gamma})$ using (9.19) one obtains

$$\operatorname{card}\{j \mid \operatorname{ht}(\tilde{\mathcal{B}}_j) \geq i + 1\} = \sum_{k=i+1}^{m+1} n_k = N_i . \tag{9.22}$$

The remaining assertions are immediate. □

PROOF OF PROPOSITION 9.4.3. By virtue of Theorem 5.2.2 one may assume that X_h satisfies (9.4) in Proposition 9.3.6 with $j = m + 1 = \operatorname{ht}(\Gamma_h) = \operatorname{ht}(\hat{\Gamma})$. Due to Remark 9.4.5 one may suppose as well that Γ_h do not have branches of height 1. By virtue of Lemma 9.4.7 the assertion of Proposition 9.3.6 follows from the next claim by letting $i = m + 1$.

CLAIM. For any $i = 1, \dots, m+1$ there is an isomorphism $V_i \cong_B \mathcal{Y}_i$ which sends the components of $z^*(0)$ in V_i to components of $z^*(0)$ in $\mathcal{Y}_i$ preserving the levels.

PROOF OF THE CLAIM. We proceed by induction on i using the trivializing sequences (9.18) and (9.21). By our assumption for $i = 1$ one has $N_1 = N_0 = N$, see (9.22). Recall (see Lemma 9.4.9) that $V_1 \subset \mathbb{A}^5 = \operatorname{Spec} \Bbbk[z, u, v, t, w]$ is given by

$$zt - p_1(u) = zw - q(u, v) = 0\,.$$

Thus, V_1 results from an affine modification $V_1 \to \mathbb{A}^3 = \operatorname{Spec} \Bbbk[z, u, v]$ along divisor $z^*(0)$ whose center is the ideal $(z, p_1(u), q(u, v)) \subset \Bbbk[z, u, v]$ supported by a reduced zero dimensional scheme $\mathfrak{S}_0 \subset \{z = 0\}$ of cardinality N. By Lemma 5.1.1 the cylinder $\mathcal{Y}_1$ results as well from an affine modification $\mathcal{Y}_1 \to \mathbb{A}^3 = B \times \mathbb{A}^2 = \operatorname{Spec} \Bbbk[z, u, v]$ along divisor $z^*(0)$ whose center is a reduced zero dimensional scheme $\mathcal{S}_0 \subset \{z = 0\}$ of cardinality N. There is an automorphism τ of $\mathbb{A}^3$ with $\tau_*(z) = z$ and $\tau(\mathfrak{S}_0) = \mathcal{S}_0$. By Lemma 1.4.3, τ can be lifted to an isomorphism $\varphi_1 \colon V_1 \xrightarrow{\cong_B} \mathcal{Y}_1$. Thus, the claim holds for $i = 1$.

Suppose further that for some $i \in \{1, \dots, m\}$ there is an isomorphism $\varphi_i \colon V_i \xrightarrow{\cong_B} \mathcal{Y}_i$ which transforms the N affine planes $\mathcal{F}_{i,(\alpha,\beta)}$ in (9.17) to the N affine planes $\tilde{\mathcal{F}}_{i,j} \subset \mathcal{Y}_i$ preserving the levels, cf. Lemmas 9.4.10 and 9.4.11. The N_i planes $\mathcal{F}_{i,(\alpha,\beta)} \subset V_i$ which carry the center $\mathfrak{S}_i = \cup_{l_h(\alpha) \ge i+1} L_{i,(\alpha,\beta)}$ of the blowup $\nu_{i+1} \colon V_{i+1} \to V_i$ are situated on the top level i. Hence their images $\tilde{\mathcal{F}}_{i,j}$ are as well components of $z^*(0)$ in $\mathcal{Y}_i$ on the top level i. These N_i affine planes $\tilde{\mathcal{F}}_{i,j}$ do not coincide, in general, with the N_i top level components of $z^*(0)$ in $\mathcal{Y}_i$ which carry the center $\mathcal{S}_i$ of the affine modification $s_{i+1} \colon \mathcal{Y}_{i+1} \to \mathcal{Y}_i$ from (9.21). However, there is an automorphism T of the bush $\Gamma_0(\pi_{Y_i}) = (\tilde{\Gamma})_{\le i}$ sending the N_i branches of height i carrying the vertices which correspond to the image of $\mathfrak{S}_i$ to the N_i branches of height i carrying the vertices which correspond to $\mathcal{S}_i$. By Theorem 5.2.2, T is induced by an automorphism $\tilde{\tau} \in \operatorname{SAut}_B(\mathcal{Y}_i)$. Applying $\tilde{\tau}$ one may suppose that the same collection $\mathfrak{F}_i$ of top level i components of $z^*(0)$ in $\mathcal{Y}_i$ supports the center $\mathcal{S}_i$ of modification s_{i+1} and the image of the center $\mathfrak{S}_i$ of ν_{i+1}. Now each component $\tilde{\mathcal{F}}_{i,j} \in \mathfrak{F}_i$ carries two distinct embedded affine lines contained in $\mathfrak{S}_i$ and $\mathcal{S}_i$, respectively. By the relative Abhyankar-Moh-Suzuki Theorem (see Proposition 4.4.1) there exists an automorphism $\eta_i \in \operatorname{SAut}_B(\mathcal{Y}_i)$ which sends the image of $\mathfrak{S}_i$ to $\mathcal{S}_i$ and preserves the levels of the components of $z^*(0)$. Composing φ_i and η_i one may suppose that $\varphi_i(\mathfrak{S}_i) = \mathcal{S}_i$ while φ_i is still level-preserving. By Lemma 1.4.3, φ_i admits a lift to a level-preserving isomorphism $\varphi_{i+1} \colon V_{i+1} \xrightarrow{\cong_B} \mathcal{Y}_{i+1}$ which fits in a commutative diagram

$$\begin{array}{ccc} V_{i+1} & \xrightarrow[\cong_B]{\varphi_{i+1}} & \mathcal{Y}_{i+1} \\ {\scriptstyle \nu_{i+1}}\big\downarrow & & \big\downarrow{\scriptstyle s_{i+1}} \\ V_i & \xrightarrow[\varphi_i]{\cong_B} & \mathcal{Y}_i \end{array} \tag{9.23}$$

Passing from V_i to V_{i+1} increases by 1 the levels of the components of $z^*(0)$ which carry the center of the blowup while the levels of the remaining components do not change. The same is true for the passage from $\mathcal{Y}_i$ to $\mathcal{Y}_{i+1}$. It follows that φ_{i+1} still preserves the levels. This concludes the inductive step. □

9.5. Cylinders over Danielewski-Fieseler surfaces

In this section we restrict to the GDF surfaces with a unique special fiber. Our main result (Theorem 9.5.3) says that, up to isomorphism of cylinders over the base, it suffices to consider only the surfaces whose fiber tree of the unique special fiber is a bush.

DEFINITION 9.5.1. We say that a marked GDF surface $\pi: X \to B$ over a smooth affine curve B with a marking $z \in \mathcal{O}_B(B) \setminus \{0\}$ is a *marked Danielewski-Fieseler surface* if

- $z^*(0)$ is a reduced divisor supported at a single point $b_0 \in B$;
- the restriction $\pi|_{X \setminus \pi^{-1}(b_0)}: X \setminus \pi^{-1}(b_0) \to B \setminus \{b_0\}$ is a trivial line bundle.

Under these assumptions there exists a trivializing sequence (2.7) such that for any $i = 1, \ldots, m-1$ the divisor of the fibered modification $\varrho_{i+1}: X_{i+1} \to X_i$ is $z^*(0)$ on X_i.

REMARK 9.5.2. The proofs of Propositions 9.3.6 and 9.4.3 go verbatim after replacing everywhere the Danielewski-Fieseler surfaces over the pair $(\mathbb{A}^1, 0)$ by the marked Danielewski-Fieseler surfaces over the pair (B, b_0) and replacing every affine space $\mathbb{A}^s$ by the product $B \times \mathbb{A}^{s-1}$.

The main result in this section is the following.

THEOREM 9.5.3. *Let $\pi_X: X \to B$ be a marked Danielewski-Fieseler surface as in Definition 9.5.1. Then there is another marked Danielewski-Fieseler surface $\pi_Y: Y \to B$ with the same marking such that*

- *the fiber tree $\Gamma_{b_0}(\pi_Y)$ is a bush with $\operatorname{tp}(\Gamma_{b_0}(\pi_Y)) = \operatorname{tp}(\Gamma_{b_0}(\pi_X))$;*
- *there is an isomorphism over B of cylinders $\mathcal{X} \xrightarrow{\cong_B} \mathcal{Y}$ which sends the components of $\pi_{\mathcal{X}}^{-1}(b_0)$ to components of $\pi_{\mathcal{Y}}^{-1}(b_0)$ preserving the levels.*

PROOF. Consider a trivializing sequence (2.7) for $X = X_m$ with $\Gamma_{b_0}(\pi_{X_i}) = \Gamma_{\le i}$ where $\Gamma = \Gamma_{b_0}(\pi_X)$. There is a unique, up to isomorphism, bush $\operatorname{bh}(\Gamma)$ such that $\operatorname{tp}(\operatorname{bh}(\Gamma)) = \operatorname{tp}(\Gamma)$.

If $m \le 1$ then Γ is already a bush. Suppose further that $m \ge 2$. Assume by induction that the assertion holds for X_{m-1}, that is, there is a marked Danielewski-Fieseler surface $\pi_{Y_{m-1}}: Y_{m-1} \to B$ sharing the same marking $z \in \mathcal{O}_B(B)$ with $\pi_X: X \to B$ where $z^*(0) = b_0$ such that

- $\Gamma_{b_0}(\pi_{Y_{m-1}}) = \operatorname{bh}(\Gamma_{\le m-1})$ [1] and
- there is an isomorphism $\varphi_{m-1}: \mathcal{X}_{m-1} \xrightarrow{\cong_B} \mathcal{Y}_{m-1}$ preserving the levels of special fiber components.

Recall that $\varrho_m: X_m \to X_{m-1}$ is a fibered modification along the reduced divisor $z^*(0) = \sum_{i=1}^N F_i$ on X_{m-1} with a reduced center $S = \sum_{i=1}^N S_i$ where for every $i = 1, \ldots, N$ either $S_i = F_i$ or $S_i \subset F_i$ is nonempty and finite. Let $\mathfrak{F}$ be the collection of the fiber components F_i in X_{m-1} on the top level $m-1$ with finite S_i. Consider the affine modification

$$\tilde{\varrho}_m: \mathcal{X}_m \to \mathcal{X}_{m-1}, \quad \tilde{\varrho}_m = \varrho_m \times \operatorname{id}_{\mathbb{A}^1},$$

along the reduced divisor $z^*(0) = \sum_{i=1}^N \mathcal{F}_i$ on $\mathcal{X}_{m-1}$ where $\mathcal{F}_i = F_i \times \mathbb{A}^1 \cong \mathbb{A}^2$ with the reduced center $\mathcal{S} = \bigcup_{i=1}^N \mathcal{S}_i$ where $\mathcal{S}_i = S_i \times \mathbb{A}^1$, $i = 1, \ldots, N$.

[1] Notice that $\operatorname{bh}(\Gamma_{\le i}) \ne (\operatorname{bh}(\Gamma))_{\le i}$ in general.

Let $z^*(0) = \sum_{i=1}^N \mathcal{F}_i'$ on $\mathcal{Y}_{m-1}$ where $\mathcal{F}_i' = \varphi_{m-1}(\mathcal{F}_i)$, and let $\mathcal{S}_i' = \varphi_{m-1}(\mathcal{S}_i) \subset \mathcal{F}_i'$. Applying a suitable automorphism $\tau \in \mathrm{SAut}_B \mathcal{Y}_{m-1}$ one may suppose that

$$\mathcal{S}_i' = S_i' \times \mathbb{A}^1 \subset F_i' \times \{0\} \subset \mathcal{F}_i' \quad \forall F_i \in \mathfrak{F}\,. \tag{9.24}$$

Indeed, choose for any $F_i \in \mathfrak{F}$ a point $s_i \in S_i$. Due to the relative Abhyankar-Moh-Suzuki Theorem (see Proposition 4.4.1) one can rectify the images $\varphi_{m-1}(\{s_i\} \times \mathbb{A}^1) \subset \mathcal{F}_i' \cong \mathbb{A}^2$ simultaneously for all $F_i \in \mathfrak{F}$. Then $\tau \circ \varphi_{m-1}$ sends any parallel line $\{t_i\} \times \mathbb{A}^1 \subset \mathcal{F}_i = F_i \times \mathbb{A}^1$ where $t_i \in S_i \setminus \{s_i\}$ to an affine line, say, $\{t_i'\} \times \mathbb{A}^1$ parallel to (and disjoint with) the image $\{s_i'\} \times \mathbb{A}^1 = \tau \circ \varphi_{m-1}(\{s_i\} \times \mathbb{A}^1)$.

Assuming (9.24) perform the fibered modification $\varrho_m' : Y_m \to Y_{m-1}$ along the reduced divisor $z^*(0)$ on Y_{m-1} with the reduced center $\sum_{i=1}^N S_i'$. It results in a marked Danielewski-Fieseler surface $\pi_{Y_m} : Y_m \to B$. Consider the induced affine modification of cylinders

$$\tilde{\varrho}_m' : \mathcal{Y}_m \to \mathcal{Y}_{m-1}, \quad \tilde{\varrho}_m' = \varrho_m' \times \mathrm{id}_{\mathbb{A}^1}\,,$$

along the reduced divisor $z^*(0) = \sum_{i=1}^N \mathcal{F}_i'$ on $\mathcal{Y}_{m-1}$ with the reduced center $\mathcal{S}' = \varphi_{m-1}(\mathcal{S}_i) = \sum_{i=1}^N \mathcal{S}_i'$ satisfying (9.24). Since φ_{m-1} sends the divisor and the center of the affine modification $\mathcal{X}_m \to \mathcal{X}_{m-1}$ to the ones of $\mathcal{Y}_m \to \mathcal{Y}_{m-1}$ then by Lemma 1.4.3, φ_{m-1} admits a lift to an isomorphism $\varphi_m : \mathcal{X}_m \xrightarrow{\cong_B} \mathcal{Y}_m$ which automatically preserves the levels.

The fiber tree $\Gamma_{b_0}(\pi_{Y_m})$ is a spring bush with $\mathrm{tp}(\tilde{\Gamma}) = \mathrm{tp}(\Gamma)$. Due to Proposition 9.4.3 and Remark 9.5.2 there exists a marked Danielewski-Fieseler surface $\pi_Y : Y \to B$ sharing the same marking z such that

- $\Gamma_{b_0}(\pi_Y) = \mathrm{bh}(\Gamma_{b_0}(\pi_{Y_m}))$ is a bush and
- there is an isomorphism of cylinders $\mathcal{Y} \xrightarrow{\cong_B} \mathcal{Y}_m$.

Since $\mathcal{X} = \mathcal{X}_m \cong_B \mathcal{Y}_m$ under φ_m it follows that

- $\mathrm{tp}(\Gamma_{b_0}(\pi_Y)) = \mathrm{tp}(\Gamma)$ and
- there is an isomorphism of cylinders $\mathcal{X} \xrightarrow{\cong_B} \mathcal{Y}$,

as required. □

9.6. Proof of the main theorem

9.6.1. Invariance of the type up to a linear equivalence. The 'only if' part of Theorem 9.2.1 follows from the next proposition.

PROPOSITION 9.6.1. *Consider two GDF surfaces $\pi_X : X \to B$ and $\pi_Y : Y \to B$ over the same base B. Assume that there is an isomorphism over B of cylinders $\varphi : \mathcal{X} \xrightarrow{\cong_B} \mathcal{Y}$. Let $\tau : \mathrm{DF}(\pi_X) \xrightarrow{\cong_B} \mathrm{DF}(\pi_Y)$ be the induced isomorphism of the Danielewski-Fieseler quotients, see Lemma 9.2.2. Then the type divisors $\mathrm{tp.div}(\pi_X)$ and $\tau^*(\mathrm{tp.div}(\pi_Y))$ on $\mathrm{DF}(\pi_X)$ are linearly equivalent.*

PROOF. Consider the birational morphisms

$$\sigma_X : X \to X_0 = B \times \mathbb{A}^1 \quad \text{and} \quad \sigma_Y : Y \to Y_0 = B \times \mathbb{A}^1$$

which are compositions of the contractions ϱ_i in (2.7). Consider also the induced birational morphisms of cylinders

$$\sigma_{\mathcal{X}} : \mathcal{X} \to B \times \mathbb{A}^2,\ \sigma_{\mathcal{X}} = \sigma_X \times \mathrm{id}_{\mathbb{A}^1}, \quad \text{and} \quad \sigma_{\mathcal{Y}} : \mathcal{Y} \to B \times \mathbb{A}^2,\ \sigma_{\mathcal{Y}} = \sigma_Y \times \mathrm{id}_{\mathbb{A}^1}\,. \tag{9.25}$$

There is a birational map $\psi: B \times \mathbb{A}^2 \dashrightarrow B \times \mathbb{A}^2$ fitting in the commutative diagram

$$(9.26)\qquad \begin{array}{ccc} \mathcal{X} & \xrightarrow[\cong_B]{\varphi} & \mathcal{Y} \\ \downarrow \sigma_{\mathcal{X}} & & \downarrow \sigma_{\mathcal{Y}} \\ B \times \mathbb{A}^2 & \overset{\psi}{\dashrightarrow} & B \times \mathbb{A}^2 . \end{array}$$

Fix a common marking $z \in \mathcal{O}_B(B)$ with $z^*(0) = b_1 + \ldots + b_n$ for $\pi_X: X \to B$ and for $\pi_Y: Y \to B$. Letting $B^* = B \setminus \{b_1, \ldots, b_n\} \subset B$, ψ restricts to an automorphism over B^* of $B^* \times \mathbb{A}^2$. Letting $\mathbb{A}^2 = \operatorname{Spec} \Bbbk[u,v]$ the Jacobian $J(\psi)(b)$ of $\psi|_{\{b\} \times \mathbb{A}^2} \in \operatorname{Aut} \mathbb{A}^2$ is a rational function on B without zeros and poles in B^*.

CLAIM. Consider a fiber component $\mathcal{F} = F \times \mathbb{A}^1 \subset \pi_{\mathcal{X}}^{-1}(b_i)$ and its image $\mathcal{F}' = \varphi(\mathcal{F}) \subset \pi_{\mathcal{Y}}^{-1}(b_i)$. Let l and l' be the levels of F and F', respectively. Then one has

$$(9.27)\qquad \operatorname{ord}_{b_i} J(\psi) = l' - l .$$

This order does not depend on the choice of a component $\mathcal{F}$ of $\pi_{\mathcal{X}}^{-1}(b_i)$.

PROOF OF THE CLAIM. Let $\mathcal{P}$ ($\mathcal{P}'$, respectively) be the path of length l (l', respectively) in $\Gamma_{b_i}(\pi_X)$ ($\Gamma_{b_i}(\pi_Y)$, respectively) joining the leaf $\bar{F}$ ($\bar{F}'$, respectively) with the root. The chains $\mathcal{P}$ and $\mathcal{P}'$ are the dual graphs of the total transforms of $\{b_i\} \times \mathbb{P}^1 \subset \bar{B} \times \mathbb{P}^1$ in (2.8) under certain sequences of blowups with centers in some points $\alpha, \alpha' \in \{b_i\} \times \mathbb{A}^1$, respectively, and infinitely near points.

Choose local coordinates (z,u) near the fiber $\{b_i\} \times \mathbb{A}^1$ in $B_i \times \mathbb{A}^1$ so that $\alpha = (0,0) \in \{b_i\} \times \mathbb{A}^1$. The chain of affine modifications which corresponds to $\mathcal{P}$ yields in suitable local coordinates the chain of extensions

$$(9.28)\qquad \Bbbk[z,u] \subset \Bbbk[z,u/z] \subset \Bbbk[z,u/z^2] \subset \ldots \subset \Bbbk[z,u/z^l] .$$

Letting $B_i = B^* \cup \{b_i\}$ consider the standard neighborhoods $U_F \cong_{B_i} B_i \times \mathbb{A}^1$ of F in X and $U_{F'} \cong_{B_i} B_i \times \mathbb{A}^1$ of F' in Y, see Proposition 3.2.2, along with their cylinders

$$\mathcal{U}_{\mathcal{F}} = U_F \times \mathbb{A}^1 \subset \mathcal{X} \quad \text{and} \quad \mathcal{U}_{\mathcal{F}'} = U_{F'} \times \mathbb{A}^1 \subset \mathcal{X}' \quad \text{where} \quad \mathcal{U}_{\mathcal{F}} \cong_{B_i} \mathcal{U}_{\mathcal{F}'} \cong_{B_i} B_i \times \mathbb{A}^2 .$$

Let $\Omega_i = B_i \times \mathbb{A}^2$ be the standard neighborhood of $\{b_i\} \times \mathbb{A}^2$ in $B_i \times \mathbb{A}^2$. Due to (9.28) the restriction $\sigma_{\mathcal{X}}|_{\mathcal{U}_{\mathcal{F}}}: \mathcal{U}_{\mathcal{F}} \to \Omega_i$ can be given in suitable local coordinates (z,t,v) in $\mathcal{U}_{\mathcal{F}}$ near $\mathcal{F}$ and (z,u,v) in Ω_i near $\{b_i\} \times \mathbb{A}^2$ as

$$\sigma_{\mathcal{X}}: (z,t,v) \mapsto (z, z^l t, v) ,$$

cf. Corollary 9.3.8(a). Similarly, the restriction $\sigma_{\mathcal{Y}}|_{\mathcal{U}_{\mathcal{F}'}}: \mathcal{U}_{\mathcal{F}'} \to \Omega_i'$ can be given in suitable local coordinates as

$$\sigma_{\mathcal{Y}}: (z,t',v') \mapsto (z, z^{l'} t', v') .$$

In these local coordinates one obtains

$$\det(d_{(t,v)} \sigma_{\mathcal{X}}) = z^l \quad \text{and} \quad \det(d_{(t',v')} \sigma_{\mathcal{Y}}) = z^{l'} .$$

Since $\det(d_{(t,v)} \varphi)$ is an invertible function on B one has $J(\psi)(z) \sim z^{l'-l}$ near b_i. This yields (9.27). Now the claim follows.

Using (9.27) one concludes that

$$(9.29)\qquad \tau^*(\operatorname{tp.div}(\pi_Y)) - \operatorname{tp.div}(\pi_X) = p^*(\operatorname{div} J(\psi))$$

where $p: \operatorname{DF}(\pi_X) \to B$ is the natural projection. Now the proof is completed. □

The following corollary is immediate from (9.29).

COROLLARY 9.6.2. *Under the assumptions of Proposition* 9.6.1, $\psi \in \mathrm{Aut}_B(B \times \mathbb{A}^2)$ *if and only if* $\mathrm{tp.div}(\pi_X) = \tau^*(\mathrm{tp.div}(\pi_Y))$.

9.6.2. Special isomorphisms and regularization. In this section we finish the proof of Theorem 9.2.1. We need the following notions.

DEFINITION 9.6.3 (*Special isomorphisms*). Given marked GDF surfaces $\pi_X: X \to B$ and $\pi_Y: Y \to B$ sharing a common marking $z \in \mathcal{O}_B(B) \setminus \{0\}$ consider trivializing sequences of cylinders (9.12) along with the corresponding birational morphisms $\sigma_{\mathcal{X}}$ and $\sigma_{\mathcal{Y}}$ as in (9.25). Given an isomorphism $\varphi: \mathcal{X} \xrightarrow{\cong_B} \mathcal{Y}$ consider a birational B-automorphism $\psi = \sigma_{\mathcal{Y}} \circ \varphi \circ \sigma_{\mathcal{X}}^{-1}$ of $B \times \mathbb{A}^2$ biregular off $z^{-1}(0)$ and fitting in diagram (9.26). We say that φ is *special* if $\psi \in \mathrm{SAut}_{B^*}(B^* \times \mathbb{A}^2)$ where $B^* = B \setminus z^{-1}(0)$ and *sub-special* if $\psi \in \mathrm{SAut}_{B'}(B' \times \mathbb{A}^2)$ for some Zariski open dense subset $B' \subset B$.

One has the following criterion for an isomorphism to be sub-special.

LEMMA 9.6.4. *An isomorphism* $\varphi: \mathcal{X} \xrightarrow{\cong_B} \mathcal{Y}$ *is sub-special if and only if* ψ *in* (9.26) *verifies*

$$J(\psi)(b) = \mathrm{Jac}(\psi|_{\{b\} \times \mathbb{A}^2}) = 1 \quad \forall b \in B^* . \tag{9.30}$$

If (9.30) *holds then one has a factorization*

$$\psi = \prod_{i=1}^{N} \exp(\partial_i) \tag{9.31}$$

where $\partial_1, \ldots, \partial_N$ *are locally nilpotent derivations of* $\mathcal{O}_{B'}(B')[u,v]$ *for an open dense subset* $B' \subset B$.

PROOF. The 'only if' part of the first assertion is well known, see, e.g., [**2**, Lem. 4.10]. Assume further that ψ satisfies (9.30). Let $\mathcal{B} = \mathcal{O}_{B^*}(B^*)$, and let $K = \mathrm{Frac}\, \mathcal{O}_B(B)$ be the function field of B. Since $\psi \in \mathrm{Aut}_{\mathcal{B}}\, \mathcal{B}[u,v]$ and $J(\psi) = 1$ one has $\psi \in \mathrm{SAut}_K\, K[u,v]$, see [**37**, Prop. 9] or [**62**, Example 2.1]. Note that the base field K in [**37**] and [**62**] is supposed to be algebraically closed of characteristic zero. However, due to the van der Kulk version of the Jung Theorem (see, e.g., [**70**]) the algebraic closeness assumption is superfluous.

The group $\mathrm{SAut}_K\, K[u,v]$ is generated by the replicas of the locally nilpotent derivations $\partial/\partial u$ and $\partial/\partial v$ ([**62**, Example 2.1]). Therefore, one has a decomposition

$$\psi = \prod_{i=1}^{m} \exp(f_i \partial/\partial u) \exp(g_i \partial/\partial v) = \prod_{j=1}^{N} \exp(\partial_j), \quad N = 2m\,, \tag{9.32}$$

where $f_i \in K[v]$, $g_i \in K[u]$, and the ∂_j are locally nilpotent K-derivations of $K[u,v]$, $i = 1, \ldots, n$, $j = 1, \ldots, N$. This yields (9.31). □

A regularization procedure described below allows to replace certain isomorphism of cylinders by special ones.

PROPOSITION 9.6.5. *Under the assumptions of Proposition* 9.6.1 *suppose in addition that* $\mathrm{tp.div}(\pi_X) = \tau^*(\mathrm{tp.div}(\pi_Y))$. *Then there exists a special isomorphism* $\tilde{\varphi}: \mathcal{X} \xrightarrow{\cong_B} \mathcal{Y}$.

PROOF. Due to Corollary 9.6.2, ψ in (9.26) is biregular, and so, the Jacobian $J(\psi)(b)$ is a non-vanishing regular function on B. Consider the automorphism

$$\eta \in \mathrm{Aut}_B \mathcal{Y}, \qquad \eta: (y,v) \mapsto (y, v/J(\psi)) \quad \text{for} \quad (y,v) \in \mathcal{Y} = Y \times \mathbb{A}^1 .$$

Then $\hat{\varphi} := \eta \circ \varphi: \mathcal{X} \xrightarrow{\cong_B} \mathcal{Y}$ and $\hat{\psi} := \sigma_{\mathcal{Y}} \circ \hat{\varphi} \circ \sigma_{\mathcal{X}}^{-1} \in \mathrm{Aut}(B \times \mathbb{A}^2)$ fit in (9.26). One has $J(\hat{\psi}) \equiv 1$. By Lemma 9.6.4, $\hat{\varphi}$ is sub-special. To simplify the notation we will suppose that φ is.

According to Lemma 9.6.4, ψ can be factorized as in (9.32). For a natural $s \gg 1$ choose a function $h \in \mathcal{O}_B(B) \smallsetminus \{0\}$ such that

- $hf_i \in \mathcal{O}_{B^*}(B^*)[u]$ and $hg_i \in \mathcal{O}_{B^*}(B^*)[v]$ where f_i, g_i are as in (9.32), $i = 1, \ldots, m$;
- $h - 1$ vanishes to order $s + M$ at $b_1, \ldots, b_n$ where M is the maximal order of pole at $b_1, \ldots, b_n$ of the coefficients of f_i, g_i, $i = 1, \ldots, m$.

Replacing the factors of ψ in (9.32) by their h-replicas yields an automorphism

$$\psi_h = \prod_{i=1}^{N} \exp(h\partial_i) \in \mathrm{SAut}_{B^*}(B^* \times \mathbb{A}^2), \qquad N = 2m . \tag{9.33}$$

CLAIM. Letting $\alpha = \psi_h \circ \psi^{-1}$ one has $\alpha \equiv \mathrm{id} \mod z^s$ and $\alpha^{-1} \equiv \mathrm{id} \mod z^s$.

PROOF OF THE CLAIM. We prove the first congruence, the proof of the second one being similar. Let

$$\alpha_j = \left(\prod_{i=1}^{j} \exp(h\partial_i)\right)\left(\prod_{i=1}^{j} \exp(\partial_i)\right)^{-1} = \left(\prod_{i=1}^{j-1} \exp(h\partial_i)\right)\exp((h-1)\partial_j)\left(\prod_{i=1}^{j-1} \exp(\partial_i)\right)^{-1} .$$

Since $(h-1)\partial_j \equiv 0 \mod z^s$ one has

$$\exp((h-1)\partial_j) - \mathrm{id} \equiv 0 \mod z^s .$$

Hence

$$\alpha_1 \equiv \mathrm{id} \mod z^s \quad \text{and} \quad \alpha_j - \alpha_{j-1} \equiv 0 \mod z^s .$$

By recursion one gets $\alpha = \alpha_N \equiv \mathrm{id} \mod z^s$, as needed.

Notice that α satisfies the conditions of Lemma 1.4.4 with respect to the affine modification $\sigma_{\mathcal{Y}}: \mathcal{Y} \to B \times \mathbb{A}^2$ along divisor $(z^t)^*(0)$ where $t = \mathrm{ht}(\mathcal{D}(\pi_X))$. Indeed, $\alpha_*(z) = z$, and $\alpha \equiv \mathrm{id} \equiv \alpha^{-1} \mod z^{t\lfloor s/t \rfloor}$ where $s \gg t$. By Lemma 1.4.4, $\alpha \in \mathrm{SAut}_B(B \times \mathbb{A}^2)$ admits a lift $\tilde{\alpha} \in \mathrm{Aut}_B \mathcal{Y}$ such that $\tilde{\alpha} \equiv \mathrm{id} \mod z^{t\lfloor s/t \rfloor - t}$. Letting $\tilde{\varphi} = \tilde{\alpha} \circ \varphi: \mathcal{X} \to \mathcal{Y}$ and $\tilde{\psi} = \alpha \circ \psi = \psi_h \in \mathrm{SAut}_{B^*}(B^* \times \mathbb{A}^2)$ yields a pair $(\tilde{\varphi}, \tilde{\psi})$ fitting in (9.26). So, the isomorphism $\tilde{\varphi}: \mathcal{X} \xrightarrow{\cong_B} \mathcal{Y}$ is special. □

PROOF OF THE 'IF' PART OF THEOREM 9.2.1. It is worth to start by recalling the setup. We consider two marked GDF surfaces $\pi_X: X \to B$ and $\pi_Y: Y \to B$ over the same base B and with the same marking $z \in \mathcal{O}_B(B)$ where $z^*(0) = b_1 + \ldots + b_n$. We assume that there is an isomorphism $\tau: \mathrm{DF}(\pi_X) \xrightarrow{\cong_B} \mathrm{DF}(\pi_Y)$ such that $\mathrm{tp.div}(\pi_X) \sim \tau^*(\mathrm{tp.div}(\pi_Y))$ on $\mathrm{DF}(\pi_X)$. We must show that under these assumptions there exists an isomorphism $\varphi: \mathcal{X} \xrightarrow{\cong_B} \mathcal{Y}$.

By Lemma 7.1.11 one may suppose that

$$\mathrm{tp.div}(\pi_X) = \tau^*(\mathrm{tp.div}(\pi_Y)) . \tag{9.34}$$

We proceed by induction on the number $n = \mathrm{card}\, z^{-1}(0)$ of special fibers. If $n = 1$ then $\pi_X: X \to B$ and $\pi_Y: Y \to B$ are marked Danielewski-Fieseler surfaces as in

Definition 9.5.1. By Theorem 9.5.3 there exist two other marked Danielewski-Fieseler surfaces $\pi_{X'}: X' \to B$ and $\pi_{Y'}: Y' \to B$ with the same marking such that

- there are isomorphisms $\mathcal{X}' \cong_B \mathcal{X}$ and $\mathcal{Y}' \cong_B \mathcal{Y}$, and
- the fiber trees $\Gamma_{b_1}(\pi_{X'})$ and $\Gamma_{b_1}(\pi_{Y'})$ are bushes with

$$\mathrm{tp}(\Gamma_{b_1}(\pi_{X'})) = \mathrm{tp}(\Gamma_{b_1}(\pi_X)) \quad \text{and} \quad \mathrm{tp}(\Gamma_{b_1}(\pi_{Y'})) = \mathrm{tp}(\Gamma_{b_1}(\pi_Y))\,.$$

Then (9.34) implies

$$\mathrm{tp}(\Gamma_{b_1}(\pi_{X'})) = \mathrm{tp}(\Gamma_{b_1}(\pi_{Y'}))\,.$$

Therefore, the bushes $\Gamma_{b_1}(\pi_{X'})$ and $\Gamma_{b_1}(\pi_{Y'})$ are isomorphic. By Theorem 5.2.2 there is an isomorphism of cylinders $\mathcal{X}' \cong_B \mathcal{Y}'$, hence also an isomorphism $\varphi: \mathcal{X} \xrightarrow{\cong_B} \mathcal{Y}$. This proves the assertion for $n = 1$.

Suppose the assertion holds if $\operatorname{card} z^{-1}(0) \le n - 1$. Consider further the case $\operatorname{card} z^{-1}(0) = n$. Let

$$B_0 = B \setminus \{b_2, \dots, b_n\} \quad \text{and} \quad B_1 = B \setminus \{b_1\} \quad \text{so that} \quad B_0 \cup B_1 = B \quad \text{and} \quad B_0 \cap B_1 = B^*\,.$$

Let $\pi_{X_0}: X_0 \to B_0$ and $\pi_{X_1}: X_1 \to B_1$ be the restrictions of π_X and π_Y over B_0 and B_1, respectively. Define in a similar way $\pi_{Y_0}: Y_0 \to B_0$ and $\pi_{Y_1}: Y_1 \to B_1$. By the inductive hypothesis there are commutative diagrams (cf. (9.26))

$$\text{(9.35)} \qquad \begin{array}{ccc} \mathcal{X}_0 & \xrightarrow[\cong_{B_0}]{\varphi_0} & \mathcal{Y}_0 \\ {\scriptstyle \sigma_{\mathcal{X}}|\mathcal{X}_0}\downarrow & & \downarrow{\scriptstyle \sigma_{\mathcal{Y}}|\mathcal{Y}_0} \\ B_0 \times \mathbb{A}^2 & \overset{\psi_0}{\dashrightarrow} & B_0 \times \mathbb{A}^2 \end{array} \qquad \text{and} \qquad \begin{array}{ccc} \mathcal{X}_1 & \xrightarrow[\cong_{B_1}]{\varphi_1} & \mathcal{Y}_1 \\ {\scriptstyle \sigma_{\mathcal{X}}|\mathcal{X}_1}\downarrow & & \downarrow{\scriptstyle \sigma_{\mathcal{Y}}|\mathcal{Y}_1} \\ B_1 \times \mathbb{A}^2 & \overset{\psi_1}{\dashrightarrow} & B_1 \times \mathbb{A}^2 \end{array}$$

By virtue of Proposition 9.6.5 one may suppose that both φ_0 and φ_1 in (9.35) are special, see Definition 9.6.3. To cook up an isomorphism φ over B using φ_0 and φ_1 we apply the same kind of regularization as in the proof of Proposition 9.6.5.

Claim 1. Given $s \gg 1$ there exist $\psi_0', \psi_1' \in \mathrm{SAut}_B(B \times \mathbb{A}^2)$ satisfying

(i) $\psi_0' \equiv \psi_0 \mod z^s$ near $\{b_1\} \times \mathbb{A}^2$ and $\psi_0' \equiv \mathrm{id} \mod z^s$ near $\{b_i\} \times \mathbb{A}^2$, $i = 2, \dots, n$;

(ii) $\psi_1' \equiv \mathrm{id} \mod z^s$ near $\{b_1\} \times \mathbb{A}^2$ and $\psi_1' \equiv \psi_1 \mod z^s$ near $\{b_i\} \times \mathbb{A}^2$, $i = 2, \dots, n$.

Proof of Claim 1. It suffices to prove (i), the proof of (ii) being similar. Likewise in the proof of Proposition 9.6.5 we replace the factors $\exp(\partial_j)$ in the factorization $\psi_0 = \prod_j \exp(\partial_j) \in \mathrm{SAut}_{B^*}(B^* \times \mathbb{A}^2)$ as in (9.33) by their replicas $\exp(h\partial_j)$ where $h \in \mathcal{O}_B(B)$ satisfies

- $h - 1$ vanishes to order $s' \gg s$ at b_1;
- h vanishes to order s at $b_2, \dots, b_n$.

This gives an automorphism $\psi_0' = \prod_j \exp(h\partial_j) \in \mathrm{SAut}_B(B \times \mathbb{A}^2)$ verifying (i), as desired.

The following claim ends the proof of the theorem.

Claim 2. The composition $\psi = \psi_0' \circ \psi_1' \in \mathrm{SAut}_B(B \times \mathbb{A}^2)$ admits a lift to an isomorphism $\varphi: \mathcal{X} \xrightarrow{\cong_B} \mathcal{Y}$.

Proof of Claim 2. The affine modifications $\sigma_{\mathcal{X}}: \mathcal{X} \to B \times \mathbb{A}^2$ and $\sigma_{\mathcal{Y}}: \mathcal{Y} \to B \times \mathbb{A}^2$ along divisors $z^*(0)$ on $\mathcal{X}$ and $\mathcal{Y}$, respectively, have for their centers certain subschemas $\mathcal{S}_{\mathcal{X}}, \mathcal{S}_{\mathcal{Y}}$ of the rth infinitesimal neighborhood of the divisor $z^{-1}(0)$ on $B \times \mathbb{A}^2$ where $r \geq 1$ is such that z^r belongs to the corresponding defining ideals $I(\mathcal{S}_{\mathcal{X}}), I(\mathcal{S}_{\mathcal{Y}}) \subset \mathcal{O}_{B\times\mathbb{A}^2}(B\times\mathbb{A}^2)$. By Lemma 1.4.3, ψ can be lifted to an isomorphism $\varphi: \mathcal{X} \xrightarrow{\cong_B} \mathcal{Y}$ provided ψ^* sends $I(\mathcal{S}_{\mathcal{Y}})$ to $I(\mathcal{S}_{\mathcal{X}})$, or, which is equivalent, if $\psi(\mathcal{S}_{\mathcal{X}}) = \mathcal{S}_{\mathcal{Y}}$.

For $i = 0, 1$ the centers $\mathcal{S}_{\mathcal{X}_i}$ and $\mathcal{S}_{\mathcal{Y}_i}$ of the affine modifications $\sigma_{\mathcal{X}}|_{\mathcal{X}_i}$ and $\sigma_{\mathcal{Y}}|_{\mathcal{Y}_i}$ coincide with the restrictions $\mathcal{S}_{\mathcal{X}}|_{\mathcal{X}_i}$ and $\mathcal{S}_{\mathcal{Y}}|_{\mathcal{Y}_i}$, respectively. Since ψ_i admits a lift φ_i (see (9.35)) one has $\psi_i(\mathcal{S}_{\mathcal{X}_i}) = \mathcal{S}_{\mathcal{Y}_i}$, $i = 0, 1$.

Due to (i) and (ii) for $s \geq r$ one has $\psi|_{B_i\times\mathbb{A}^2} \equiv \psi_i|_{B_i\times\mathbb{A}^2} \mod z^r$ for $i = 0, 1$, that is, these automorphisms coincide in the rth infinitesimal neighborhood of $z^{-1}(0)$ in $B_i \times \mathbb{A}^2$. It follows that $\psi(\mathcal{S}_{\mathcal{X}_i}) = \mathcal{S}_{\mathcal{Y}_i}$, $i = 0, 1$. Finally one has $\psi(\mathcal{S}_{\mathcal{X}}) = \mathcal{S}_{\mathcal{Y}}$, as required. □

CHAPTER 10

On moduli spaces of GDF surfaces

We conclude the paper by constructions of a versal deformation family and an affine coarse moduli space of marked GDF surfaces with a given marking, a given cylinder, and a given graph divisor.

10.1. Coarse moduli spaces of GDF surfaces

DEFINITION 10.1.1. Consider a marked GDF surface $\pi_X: X \to B$ with a marking $z \in \mathcal{O}_B(B) \smallsetminus \{0\}$, a trivializing sequence (2.7), and a graph divisor $\mathcal{D} = \mathcal{D}(\pi_X)$. By a *family of GDF surfaces of type* $(B, z, \mathcal{D})$ we mean a pair of smooth morphisms of quasiprojective schemes $\mathfrak{X} \to \mathcal{S}$ and $\pi_{\mathfrak{X}}: \mathfrak{X} \to B$ such that

- for each point $\mathfrak{s} \in \mathcal{S}$ the fiber $X(\mathfrak{s})$ of $\mathfrak{X} \to \mathcal{S}$ over $\mathfrak{s}$ is reduced;
- the specialization $\pi(\mathfrak{s}): X(\mathfrak{s}) \to B$ over $\mathfrak{s}$ is a marked GDF surface with the marking z and the graph divisor $\mathcal{D}$;
- there is a point $\mathfrak{s}_0 \in \mathcal{S}$ such that the specialization over $\mathfrak{s}_0$ yields the initial marked GDF surface $\pi_X: X \to B$.

We say that $\mathfrak{X} \to \mathcal{S}$ is an (*affine*) *deformation family of marked GDF surfaces* if both $\mathfrak{X}$ and $\mathcal{S}$ are smooth (affine) varieties and the morphism $\mathfrak{X} \to \mathcal{S}$ is a submersion which extends to a proper deformation family of SNC completions of GDF surfaces over $\bar{B}$ such that the corresponding family of extended divisors $(D_{\text{ext}}(\mathfrak{s}))_{\mathfrak{s}\in\mathcal{S}}$ over $\mathcal{S}$ is locally trivial. This yields a locally trivial family of graph divisors $(\mathcal{D}(\mathfrak{s}) = \mathcal{D}(\pi_{X(\mathfrak{s})}))_{\mathfrak{s}\in\mathcal{S}}$. The monodromy group of the latter family is a subgroup of the finite group

$$\mathrm{Aut}_B(\mathcal{D}(\mathfrak{s}_0)) = \prod_{i=1}^{n} \mathrm{Aut}(\Gamma_i(\mathfrak{s}_0)).$$

We say that the family of graph divisors $(\mathcal{D}(\mathfrak{s}))_{\mathfrak{s}\in\mathcal{S}}$ is *trivial* if its monodromy group is.

The Grothendieck theory of moduli spaces ([**40**]) and its versions deal with proper schemes, or pairs of proper schemes. In our particular non-proper setting we adopt the following simplified definition.

DEFINITION 10.1.2. Consider a triplet $(B, z, \mathcal{D})$ where B is a smooth affine curve, $z \in \mathcal{O}_B(B) \smallsetminus \{0\}$ is a marking, and $\mathcal{D} = \sum_{i=1}^n \Gamma_i b_i$ is a graph divisor with trees $\Gamma_1, \ldots, \Gamma_n$ supported by $z^{-1}(0) = \{b_1, \ldots, b_n\} \subset B$. A scheme $\mathcal{C}(B, z, \mathcal{D})$ will be called a *coarse moduli space of GDF surfaces of type* $(B, z, \mathcal{D})$ if the following hold.

- To any marked GDF surface $\pi_X: X \to B$ of type $(B, z, \mathcal{D})$ there corresponds a unique point $\mathfrak{c}(\pi_X) \in \mathcal{C}(B, z, \mathcal{D})$, and vice versa, any point $\mathfrak{c} \in \mathcal{C}(B, z, \mathcal{D})$ corresponds to a unique, up to an isomorphism over B, marked GDF surface $\pi_{X(\mathfrak{c})}: X(\mathfrak{c}) \to B$ of type $(B, z, \mathcal{D})$;

- for any deformation family of GDF surfaces $\mathfrak{X} \to \mathcal{S}$ of type $(B, z, \mathcal{D})$ the correspondence $\mathcal{S} \to \mathcal{C}(B, z, \mathcal{D})$ sending a point $\mathfrak{s} \in \mathcal{S}$ to $\mathfrak{c}(\pi_{X(\mathfrak{s})}) \in \mathcal{C}(B, z, \mathcal{D})$ is a morphism.

Under certain restrictions there exists a coarse moduli space of GDF surfaces.

THEOREM 10.1.3. *Consider a triplet $(B, z, \mathcal{D})$ as in Definition 10.1.2. Suppose that $\mathcal{O}_B(B)^\times = \Bbbk^*$, that is, B does not admit any non-constant invertible regular function. Then the following hold.*

(a) *There exists a coarse moduli space $\mathcal{C}(B, z, \mathcal{D})$ of marked GDF surfaces of type $(B, z, \mathcal{D})$. This space $\mathcal{C}(B, z, \mathcal{D})$ is an affine variety with at most quotient singularities.*
(b) *The cylinders over any two surfaces in $\mathcal{C}(B, z, \mathcal{D})$ are isomorphic over B.*
(c) *If $\pi: X \to B$ is a marked GDF surface of type $(B, z, \mathcal{D})$ and X is not a Zariski 1-factor, that is, $\mathcal{D}(\pi_X) \cong \mathcal{D}$ is not a chain divisor then there exists a sequence of marked GDF surfaces $\pi_{X^{(k)}}: X^{(k)} \to B$ with the given marking z and the cylinders isomorphic over B to $X \times \mathbb{A}^1$ such that*

$$\dim \mathcal{C}(B, z, \mathcal{D}(\pi_{X^{(k)}})) \to \infty \quad \textit{when} \quad k \to \infty\,.$$

The proof of Theorem 10.1.3 is done at the end of the section.

10.2. The automorphism group of a GDF surface

NOTATION 10.2.1. Given an $\mathbb{A}^1$-fibered surface $\pi_X: X \to B$ we let $\mathcal{U}(\pi_X) = R_u(\mathrm{Aut}_B(X))$ be the (normal) subgroup of $\mathrm{Aut}_B(X)$ generated by all $\mathbb{G}_a$-actions on X along the fibers of π_X. For a GDF surface $\pi_X: X \to B$ with a trivializing sequence (2.7) we let $\mathbb{U}_l = \mathcal{U}(\pi_{X_l})$, $l = 1, \dots, m$ where $\mathbb{U}_m = \mathcal{U}(\pi_X)$.

The next proposition can be deduced from Theorems 6.3 and 8.24 in [**48**] and their corollaries. However, in our particular case we prefer to give a simple direct argument.

PROPOSITION 10.2.2. *Assume that $\mathcal{O}_B(B)^\times = \Bbbk^*$. Given a GDF surface $\pi_X: X \to B$ with a trivializing sequence (2.7) the following hold.*

(a) *There are natural inclusions*

$$\mathrm{Aut}_B(X) = \mathrm{Aut}_B(X_m) \subset \mathrm{Aut}_B(X_{m-1}) \subset \ldots \subset \mathrm{Aut}_B(X_0) = \mathrm{Aut}_B(B \times \mathbb{A}^1)\,. \tag{10.1}$$

(b) *One has $\mathrm{Aut}_B(B \times \mathbb{A}^1) = \mathbb{U}_0 \rtimes \mathbb{G}_m$ where $\mathbb{U}_0 \cong \mathcal{O}_B(B)$ viewed as a vector group.*
(c) *Let $m_{i,l} = \mathrm{ht}((\Gamma_i)_{\le l})$, $i = 1, \dots, n$. For any $l = 0, \dots, m$ one has*

$$\mathbb{U}_l = \mathbb{U}_0 \cap \mathrm{Aut}_B(X_l) \cong H^0(B, -D_l) \quad \textit{where} \quad D_l = \sum_{i=1}^n m_{i,l} b_i\,.$$

(d) *Let $C_{l-1} \subset X_{l-1}$ be the center of the fibered modification $\varrho_l: X_l \to X_{l-1}$ where $l \in \{1, \dots, m\}$. Then the subgroup $\mathbb{U}_l \subset \mathbb{U}_{l-1}$ acts identically on C_{l-1}.*
(e) *If $\pi_X: X \to B$ has a reducible fiber then $\mathrm{Aut}_B(X_m) \cong \mathbb{U}_m \rtimes \mu_d$ where $\mu_d \subset \mathbb{G}_m$ is a finite cyclic group.*

PROOF. (a) By Corollary 9.6.2 for any $l = 0, \dots, m$ one has a natural inclusion $\mathrm{Aut}_B(X_l) \subset \mathrm{Aut}_B(X_0)$. Now, $\varphi \in \mathrm{Aut}_B(X_{l-1})$ admits a lift to $\tilde{\varphi} \in \mathrm{Aut}_B(X_l)$ if and only if φ preserves the ideal of the center of the affine modification $\varrho_l: X_l \to X_{l-1}$. This, clearly, leads to the inclusions $\mathrm{Aut}_B(X_l) \subset \mathrm{Aut}_B(X_{l-1})$.

Statement (b) is immediate from the fact that any $\varphi \in \mathrm{Aut}_B(B \times \mathbb{A}^1)$ acts via

$$\varphi: (b, u) \mapsto (b, \alpha(b)u + \beta(b)) \quad \text{where} \quad \alpha \in \mathcal{O}_B(B)^\times = \Bbbk^* \quad \text{and} \quad \beta \in \mathcal{O}_B(B).$$

The first equality in (c) is easy and is left to the reader. The second one follows from [**48**, Thm. 6.3] (cf. also [**48**, Rem. 6.5.2]) due to the fact that locally in B near the point b_i, a locally nilpotent vertical vector field ∂ on $B \times \mathbb{A}^1$ admits a lift to X_l if and only if ∂ is of the form $\partial = f(z)z^{m_{i,l}}\partial/\partial u$ for a function f in the local ring of (B, b_i), cf. the proof of Lemma 3.1.1.

(d) The inclusion $\mathrm{Aut}_B(X_l) \subset \mathrm{Aut}_B(X_{l-1})$ in (10.1) implies that the union of the top level l fiber components in X_l is $\mathrm{Aut}_B(X_l)$-invariant. Using [**47**, Prop. 2.1] one can conclude that

$$\mathrm{Aut}_B(X_l) = \mathrm{Aut}_B(X_{l-1}, C_{l-1}).$$

Since the ind-subgroup $\mathbb{U}_l \subset \mathrm{Aut}_B(X_l)$ is connected and acts morphically on X_{l-1}, its action on the finite $\mathrm{Aut}_B(X_l)$-invariant set $C_{l-1} \subset X_{l-1}$ is identical.

(e) From (a)–(c) one obtains the inclusions

(10.2)
$$\mathrm{Aut}_B(X_l)/\mathbb{U}_l \subset \mathrm{Aut}_B(X_{l-1})/\mathbb{U}_{l-1} \subset \ldots \subset \mathrm{Aut}_B(X_0)/\mathbb{U}_0 = \mathbb{G}_m, \quad l = 0, \ldots, m.$$

Choose $l \in \{1, \ldots, m\}$ such that $\pi_{X_l}: X_l \to B$ has a reducible fiber while $\pi_{X_{l-1}}: X_{l-1} \to B$ does not. Then $\pi_{X_{l-1}}: X_{l-1} \to B$ admits a structure of a line bundle. Consider the associate $\mathbb{G}_m$-action on X_{l-1} along the fibers of $\pi_{X_{l-1}}$. Due to (10.2) one has $\mathrm{Aut}_B(X_{l-1}) = \mathbb{U}_{l-1} \rtimes \mathbb{G}_m$, and due to (10.1),

$$\mathbb{U}_l \subset \mathrm{Aut}_B(X_l) \subset \mathbb{U}_l \rtimes \mathbb{G}_m.$$

Let $\pi_{X_l}^{-1}(b_i)$ be a reducible fiber. Then the fiber $F = \pi_{X_{l-1}}^{-1}(b_i) \cong \mathbb{A}^1$ contains at least two distinct points of C_{l-1}. Let φ be an element of the $\mathbb{G}_m$-action on X_{l-1} along the fibers of $\pi_{X_{l-1}}$ which preserves the finite set $F \cap C_{l-1}$. Then φ has finite order, say, d where $d \le \mathrm{card}\,(F \cap C_{l-1})$. Since φ^d is identical on F then also $\varphi^d = \mathrm{id}$ on X_{l-1}. This shows that $\mathrm{Aut}_B(X_l) = \mathbb{U}_l \rtimes \mu_d$ where $\mu_d \subset \mathbb{G}_m$ is a finite cyclic subgroup. A similar semi-direct product decomposition holds for any subgroup $\mathrm{Aut}_B(X_i) \subset \mathrm{Aut}_B(X_l)$, $i = l, \ldots, m$, see (c) and (10.1). □

10.3. Configuration spaces and configuration invariants

NOTATION 10.3.1. Consider a marked GDF surface $\pi_X: X \to B$ with a marking $z \in \mathcal{O}_B(B)$, a trivializing sequence (2.7), and a graph divisor $\mathcal{D}(\pi_X) = \sum_{i=1}^n \Gamma_i b_i$ where $z^*(0) = b_1 + \ldots + b_n$. For a vertex $v \in \mathrm{vert}\,(\Gamma_i)$ on level $l > 0$ we let $F(v) \subset X_l$ be the corresponding top level fiber component over b_i, and let $(b_i, u(v)) \in B \times \mathbb{A}^1$ be the image of $F(v)$ under the contraction $\varrho_1 \circ \ldots \circ \varrho_l: X_l \to X_0 = B \times \mathbb{A}^1$. Fixing the shortest path $\gamma(v)$ joining v in Γ_i with the root $v_{0,i}$ of Γ_i one can construct the standard affine chart about $F(v)$ in X_l following the blowup process along the path $\gamma(v)$. Starting with $t_{v_{i,0}} = u$, where $v_{i,0}$ is the root of Γ_i, assume by recursion that t_v is already defined for a (non-extremal) vertex v of Γ_i on a level l, and let w be a vertex on the level $l+1$ joint to v by an edge. Then we let

$$t_w = (t_v - t_v(w))/z \tag{10.3}$$

where $t_v(w)$ is the t_v-coordinate of the point $\varrho_{l+1}(F(w)) \in F(v)$. By recursion one defines a system of affine coordinates on the fiber components of $z^{-1}(0)$ in X_l for all $l = 0, \dots, m$. If X_l admits a μ_d-action along the fibers of $\pi_{X_l}: X_l \to B$ then this system of affine coordinates is μ_d-invariant, cf. Lemma 3.3.2 and its proof.

DEFINITION 10.3.2 (*Configuration space of a tree*). Let Γ be a rooted tree. For a vertex $v \in \operatorname{vert}(\Gamma)$ on level l we let $r_+(v)$ be the number of incident edges joining v with vertices on level $l+1$. If $\operatorname{ht}(\Gamma) > 0$ then $r_+(v) = 0$ if and only if v is an extremal vertex different from the root v_0 of Γ. Thus, $\sum_{v \in \operatorname{vert}(\Gamma)} r_+(v)$ equals the number of edges of Γ. We let Γ^* stand for the subgraph of Γ obtained by deleting all the leaves of Γ and their incident edges.

Let $D_r \subset \mathbb{A}^r$ be the zero level of the discriminant of the universal monic polynomial $p = t^r + a_1 t^{r-1} + \dots + a_r \in \mathbb{K}[a_1, \dots, a_r][t]$ of degree $r > 0$. The open set $\mathcal{S}(r) = \mathbb{A}^r \smallsetminus D_r \subset \mathbb{A}^r$ represents the configuration space of r-points subsets (the roots of p) of $\mathbb{A}^1 = \operatorname{Spec} \Bbbk[t]$. The *configuration space of* Γ is the formal sum

$$\mathcal{S}(\Gamma) = \sum_{v \in \operatorname{vert}(\Gamma^*)} \mathcal{S}(r_+(v)) v .$$

A point $\mathfrak{s} \in \mathcal{S}(\Gamma)$ can be viewed as a collection of configurations

$$\mathfrak{s} = \{\mathfrak{s}(v) \in \mathcal{S}(r_+(v)) \,|\, v \in \operatorname{vert}(\Gamma^*)\} .$$

The underlying variety of $\mathcal{S}(\Gamma)$ is the smooth affine variety $\prod_{v \in \operatorname{vert}(\Gamma^*)} \mathcal{S}(r_+(v))$ of dimension

$$\dim \mathcal{S}(\Gamma) = \sum_{v \in \operatorname{vert}(\Gamma^*)} r_+(v) = \operatorname{card}(\operatorname{edges}(\Gamma)) .$$

The automorphism group $\operatorname{Aut}(\Gamma)$ of the rooted tree Γ acts on $\mathcal{S}(\Gamma)$ via

$$\operatorname{Aut}(\Gamma) \ni \alpha : \mathfrak{s} \mapsto \alpha_*(\mathfrak{s}) \quad \text{where} \quad \alpha_*(\mathfrak{s})(v) = \mathfrak{s}(\alpha(v)) .$$

This action is not effective, in general. Its kernel of non-effectiveness, say, K is the pointwise stabilizer of Γ^*:

$$K = \{\alpha \in \operatorname{Aut}(\Gamma) \,|\, \alpha(v) = v \ \ \forall v \in \operatorname{vert}(\Gamma^*)\} .$$

The quotient

$$\operatorname{Aut}^*(\Gamma) = \operatorname{Aut}(\Gamma)/K$$

acts effectively on $\mathcal{S}(\Gamma)$.

Consider the level l factor $\mathcal{S}_l(\Gamma)$ of $\mathcal{S}(\Gamma)$ where

$$\mathcal{S}_l(\Gamma) = \sum_{v \in \operatorname{vert}(\Gamma^*) \,|\, l(v) = l} \mathcal{S}(r_+(v)) v .$$

There is an effective action of the Abelian unipotent group $\mathbb{G}_a^h$ on $\mathcal{S}(\Gamma)$ where $h = \operatorname{ht}(\Gamma)$. It is defined as follows. For $l = 0, \dots, m-1$ the lth $\mathbb{G}_a$-factor of $\mathbb{G}_a^h$ acts identically on any component $\mathcal{S}(r_+(v))v$ with $l(v) \neq l$, and acts effectively on $\mathcal{S}_l(\Gamma)$ via the simultaneous shifts on $\tau \in \Bbbk$ of the affine coordinates in all the corresponding configurations. This $\mathbb{G}_a$-action on $\mathcal{S}_l(\Gamma)$ is free and admits a slice

$$\mathcal{S}_l^{\circ}(\Gamma) = \left\{ \mathfrak{s}_l = \sum_{v \in \operatorname{vert}(\Gamma^*) \,|\, l(v) = l} \mathfrak{s}_l(v) v \,\middle|\, \sum_{v \in \operatorname{vert}(\Gamma^*) \,|\, l(v) = l} \operatorname{barycentre}(\mathfrak{s}_l(v)) = 0 \right\} .$$

The smooth $\operatorname{Aut}^*(\Gamma)$-invariant affine variety $\mathcal{S}^{\circ}(\Gamma) \coloneqq \prod_{l=0}^{m-1} \mathcal{S}_l^{\circ}(\Gamma)$ is a slice of the free $\mathbb{G}_a^h$-action on $\mathcal{S}(\Gamma)$.

Besides, there is an effective $\mathbb{G}_m$-action on $\mathcal{S}(\Gamma)$ via the simultaneous multiplication by $\lambda \in \Bbbk^*$ of the affine coordinates in all the corresponding configurations. This action leaves the slice $\mathcal{S}^{\circ}(\Gamma)$ invariant.

The action of the semi-direct product

$$G(\Gamma) = \mathbb{G}_a^h \rtimes (\mathbb{G}_m \times \mathrm{Aut}^*(\Gamma))$$

on $\mathcal{S}(\Gamma)$ descends to an effective action of the reductive group $\mathbb{G}_m \times \mathrm{Aut}^*(\Gamma)$ on the smooth affine variety $\mathcal{S}(\Gamma)/\mathbb{G}_a^h \cong \mathcal{S}^\circ(\Gamma)$. The quotient

$$\mathfrak{M}(\Gamma) = \mathcal{S}(\Gamma)/G(\Gamma) = \mathcal{S}^\circ(\Gamma)/(\mathbb{G}_m \times \mathrm{Aut}^*(\Gamma))$$

is an affine variety with quotient singularities.

DEFINITION 10.3.3 (*Configuration space of a graph divisor*). Given a graph divisor $\mathcal{D} = \sum_{i=1}^n \Gamma_i b_i$ we define the *configuration space of* $\mathcal{D}$ to be the formal sum

$$\mathcal{S}(\mathcal{D}) = \sum_{i=1}^n \mathcal{S}(\Gamma_i) b_i .$$

The underlying smooth affine variety $\prod_{i=1}^n \mathcal{S}(\Gamma_i)$ has dimension equal to the number of edges in $\mathcal{D}$.

Letting $\mathcal{D}^* = \sum_{i=1}^n \Gamma_i^* b_i$ consider the group $\mathrm{Aut}_B^*(\mathcal{D}) = \prod_{i=1}^n \mathrm{Aut}^*(\Gamma_i)$. It acts effectively on $\mathcal{S}(\mathcal{D})$.

Let $h_i = \mathrm{ht}(\Gamma_i)$, and let $h(\mathcal{D}) = \sum_{i=1}^n h_i$. The effective action of $\mathbb{G}_a^{h_i} \rtimes \mathbb{G}_m$ on $\mathcal{S}(\Gamma_i)$ as defined in 10.3.2 extends to an effective action of $\mathbb{G}_a^{h(\mathcal{D})} \rtimes \mathbb{G}_m$ on $\mathcal{S}(\mathcal{D})$. The action of $\mathbb{G}_a^{h(\mathcal{D})}$ is free and admits a slice

$$\mathcal{S}^\circ(\mathcal{D}) = \sum_{i=1}^n \mathcal{S}^\circ(\Gamma_i) b_i .$$

The action of the affine algebraic group

$$G(\mathcal{D}) = \mathbb{G}_a^{h(\mathcal{D})} \rtimes (\mathbb{G}_m \times \mathrm{Aut}_B^*(\mathcal{D})) \tag{10.4}$$

on $\mathcal{S}(\mathcal{D})$ descends to an effective action of $\mathbb{G}_m \times \mathrm{Aut}_B^*(\mathcal{D})$ on the quotient $\mathcal{S}(\Gamma)/\mathbb{G}_a^{h(\mathcal{D})} \cong \mathcal{S}^\circ(\mathcal{D})$. The quotient

$$\mathfrak{M}(\mathcal{D}) = \mathcal{S}(\mathcal{D})/G(\mathcal{D}) = \mathcal{S}^\circ(\mathcal{D})/(\mathbb{G}_m \times \mathrm{Aut}_B^*(\mathcal{D})) \tag{10.5}$$

is an affine variety with quotient singularities.

DEFINITION 10.3.4 (*Configuration invariant*). Let $\pi_X : X \to B$ be a marked GDF surface as in 10.3.1 with the graph divisor $\mathcal{D} = \mathcal{D}(\pi_X) = \sum_{i=1}^n \Gamma_i b_i$ where $\Gamma_i = \Gamma_{b_i}(\pi_X)$. For a vertex $v \in \mathrm{vert}(\mathcal{D}^*)$ the top level l fiber component $F(v) \subset X_l$ is equipped with the natural affine coordinate t_v as in 10.3.1. Let $C_l \subset X_l$ be the center of the fibered modification $\varrho_{l+1} : X_{l+1} \to X_l$ in (2.7). We define the *configuration invariant* $\Delta(\pi_X) \in \mathcal{S}(\mathcal{D}(\pi_X))$ by the formula

$$\Delta(\pi_X)(v) = F(v) \cap C_l \in \mathcal{S}(r_+(v)), \quad v \in \mathrm{vert}(\mathcal{D}^*), \ l(v) = l .$$

If X admits a μ_d-action along the fibers of π_X then this action is well defined on X_l for any $l = 0, \dots, m$ making (2.7) equivariant. So, both the collection of centers $(C_l)_{l=0,\dots,m-1}$ and the collection of the affine coordinates t_v are invariant under μ_d. Hence the configuration invariant $\Delta(\pi_X)$ is as well.

10.4. Versal deformation families of trivializing sequences

DEFINITION 10.4.1. Consider a marked GDF surface $\pi_X: X \to B$ with a marking $z \in \mathcal{O}_B(B) \smallsetminus \{0\}$ and a trivializing sequence (2.7). By a corresponding *family of trivializing sequences* we mean a commutative diagram of varieties and morphisms

(10.6)

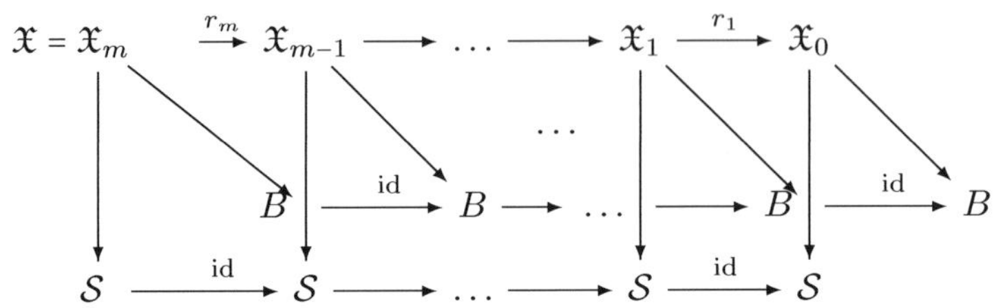

where

- $\mathfrak{X}_0 = \mathcal{S} \times (B \times \mathbb{A}^1)$ with the standard projections to the factors $\mathcal{S}$ and B;
- $\mathfrak{X}_l \to \mathcal{S}$ is a family of GDF surfaces for any $l = 0, \dots, m$;
- there is a point $\mathfrak{s}_0 \in \mathcal{S}$ such that the specialization of the upper line in (10.6) over $\mathfrak{s}_0$ yields the initial trivializing sequence $(\pi_{X_l}: X_l \to B)_{l=0,\dots,m}$.

We say that (10.6) is an *affine deformation family of trivializing sequences* if, for any $l = 0, \dots, m$, $\mathfrak{X}_l \to \mathcal{S}$ is an affine deformation family of marked GDF surfaces. It is easily seen that (10.7) extends to a diagram of trivializing families of SNC completions which specializes to (2.8) in each fiber, along with a locally trivial family of extended divisors $(D_{\text{ext}}(\mathfrak{s}))_{\mathfrak{s} \in \mathcal{S}}$.

An affine deformation family (10.6) is called *trivial* if for any $l = 1, \dots, m$ there is a commutative diagram

(10.7)

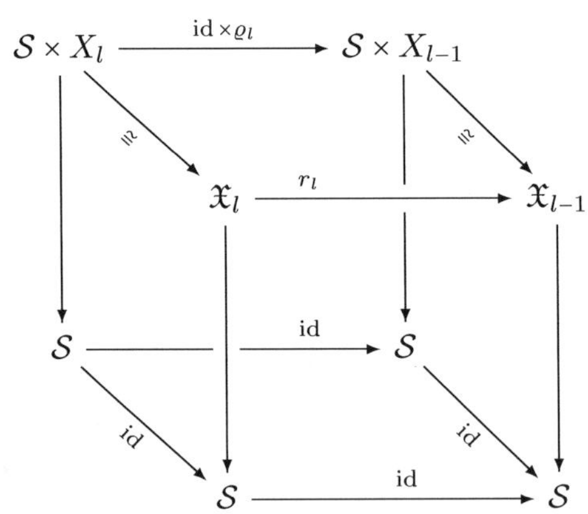

where the morphisms in the upper square are defined over B.

10.4.2 (*Deformation families of GDF surfaces over configuration spaces*). Given a triplet $(B, z, \mathcal{D})$ as in Definition 10.1.2 there is a natural deformation family

$$\mathfrak{F}(\mathcal{D}) = (\mathfrak{X}_l(\mathcal{D}) \to \mathcal{S}(\mathcal{D}))_{l=0,\dots,m}$$

of trivializing sequences (10.6) of GDF surfaces of type $(B, z, \mathcal{D})$ over the configuration space $\mathcal{S}(\mathcal{D})$ such that the associated family of graph divisors is trivial. The construction of the latter family proceeds as follows.

We let $\mathfrak{X}_0(\mathcal{D}) = \mathcal{S}(\mathcal{D}) \times (B \times \mathbb{A}^1)$ with the canonical projections to the first and the second factors. Projecting $\mathcal{S}(\mathcal{D})$ to the zero level factor

$$\mathcal{S}_0(\mathcal{D}) = \sum_{i=1}^{n} \mathcal{S}(v_{0,i}) v_{0,i} b_i$$

where $v_{0,i}$ is the root of Γ_i yields a multisection of the first projection $\mathfrak{X}_0(\mathcal{D}) \to \mathcal{S}(\mathcal{D})$. This multisection defines the center $\mathfrak{C}_0(\mathcal{D})$ of the fibered modification $r_1(\mathcal{D})\colon \mathfrak{X}_1(\mathcal{D}) \to \mathfrak{X}_0(\mathcal{D})$ fitting in diagram (10.6). Projecting now to the first level factor

$$\mathcal{S}_1(\mathcal{D}) = \sum_{i=1}^{n} \left(\sum_{v \in \mathrm{vert}\,(\Gamma_i^*) \,|\, l(v)=1} \mathcal{S}(v) v \right) b_i$$

yields the center $\mathfrak{C}_1(\mathcal{D})$ of the fibered modification $r_2(\mathcal{D})\colon \mathfrak{X}_2(\mathcal{D}) \to \mathfrak{X}_1(\mathcal{D})$ fitting in (10.6), etc. Continuing in this way one arrives finally at a deformation family $\mathfrak{F}(\mathcal{D})$ over $\mathcal{S} = \mathcal{S}(\mathcal{D})$ of trivializing sequences of marked GDF surfaces $(X_l(\mathfrak{s}))_{\mathfrak{s} \in \mathcal{S}, l=0,\dots,m}$ fitting in (10.6) with the marking z, the trivial family of graph divisors $\mathcal{D}(\pi_{X(\mathfrak{s})}) = \mathcal{D}$, and the configuration invariant $\Delta(X(\mathfrak{s})) = \mathfrak{s} \in \mathcal{S}(\mathcal{D})$ where $X(\mathfrak{s}) = X_m(\mathfrak{s})$. Clearly, it admits an extension to a family of complete surfaces with a trivial family of extended divisors.

Every morphism $f\colon \mathcal{S} \to \mathcal{S}(\mathcal{D})$ induces a family $\mathfrak{F} = f^*(\mathfrak{F}(\mathcal{D}))$ of trivializing sequences over $\mathcal{S}$ with a trivial family of graph divisors. Conversely, every such family arises in this way. This shows that $\mathfrak{F}(\mathcal{D})$ is a versal family. More precisely, the following holds.

PROPOSITION 10.4.3. *Let $\mathfrak{F} = (\mathfrak{X}_l \to \mathcal{S})_{l=0,\dots,m}$ be a family of trivializing sequences over $\mathcal{S}$. Assume that the associated family of graph divisors over $\mathcal{S}$ is trivial:*

$$\mathcal{D}(\pi_{X(\mathfrak{s})}) = \mathcal{D} \quad \forall \mathfrak{s} \in \mathcal{S}.$$

Then one has $\mathfrak{F} = \Delta^*(\mathfrak{F}(\mathcal{D}))$, that is, $\mathfrak{F}$ is induced from $\mathfrak{F}(\mathcal{D})$ via the morphism

$$\Delta\colon \mathcal{S} \to \mathcal{S}(\mathcal{D}), \quad \mathfrak{s} \mapsto \Delta(X(\mathfrak{s}))$$

defined by the configuration invariant. Consequently, the deformation family of trivializing sequences $\mathfrak{F}(\mathcal{D})$ is versal with respect to the étale topology.

PROOF. We proceed by recursion on m. The assertion is evidently true if $m = 0$. Suppose it holds for $m = l - 1$, that is, the lower square in the following diagram is commutative:

$$\begin{array}{ccc}
\mathfrak{X}_l & \xrightarrow{\varphi_l} & \mathfrak{X}_l(\mathcal{D}) \\
\downarrow r_l & & \downarrow r_l(\mathcal{D}) \\
\mathfrak{X}_{l-1} & \xrightarrow{\varphi_{l-1}} & \mathfrak{X}_{l-1}(\mathcal{D}) \\
\downarrow & & \downarrow \\
\mathcal{S} & \xrightarrow{\Delta} & \mathcal{S}(\mathcal{D})
\end{array} \qquad (10.8)$$

For $l = 0, \dots, m-1$ we let $\Delta_l\colon \mathcal{S} \to \mathcal{S}_l(\mathcal{D})$ be the composition of Δ with the projection to the level l component $\mathcal{S}_l(\mathcal{D})$ of $\mathcal{S}(\mathcal{D})$. The image of Δ_{l-1} can be seen as a multisection of $\mathfrak{X}_{l-1} \to \mathcal{S}$ which defines the center $\mathfrak{C}_{l-1} \subset \mathfrak{X}_{l-1}$ of the fibered modification $r_l\colon \mathfrak{X}_l \to \mathfrak{X}_{l-1}$. Due to diagram (10.8) one has $\mathfrak{C}_{l-1} = \varphi_{l-1}^*(\mathfrak{C}_{l-1}(\mathcal{D}))$.

According to [**47**, Prop. 2.1], φ_{l-1} admits a lift to a morphism $\varphi_l: \mathfrak{X}_l \to \mathfrak{X}_l(\mathcal{D})$ which makes the upper square in (10.8) commutative. This gives the recursive step and proves the first assertion.

To show the second one it suffices to notice that any family of graph divisors has a finite monodromy group. This monodromy becomes trivial after a suitable étale base change. According to the first part, the resulting family is induced from $\mathfrak{F}(\mathcal{D})$ via the morphism given by the configuration invariant. □

Let us study the automorphism group of $\mathfrak{F}(\mathcal{D})$ over B.

LEMMA 10.4.4. *Consider two fibers $X_l = X_l(\mathfrak{s})$ and $X_l' = X_l(\mathfrak{s}')$ of $\mathfrak{F}_l(\mathcal{D})$ where $\mathfrak{s}, \mathfrak{s}' \in \mathcal{S}(\mathcal{D})$, $l = 0, \ldots, m$. Assume that there is an isomorphism $\varphi: X = X_m \xrightarrow{\cong_B} X' = X_m'$ which induces the identity on $\mathcal{D}$. Then φ induces for any $i = 1, \ldots, n$ and $l = 0, \ldots, h_i - 1 = \operatorname{ht}(\Gamma_i^*)$ an affine transformation*

$$\varphi_{i,l}: t_v \mapsto \alpha t_v + \beta_{i,l}, \quad \alpha_i \in \Bbbk^*, \ \beta_{i,l} \in \Bbbk \tag{10.9}$$

such that $\mathfrak{s}'(v) = \varphi_{i,l}(\mathfrak{s}(v))$ for any $v \in \operatorname{vert}(\Gamma_i^*)$ with $l(v) = l$.

PROOF. Recall that φ can be extended first to an isomorphism of pseudo-minimal completions $\bar{\varphi}: \bar{X} \xrightarrow{\cong_B} \bar{X}'$, and then to an isomorphism of the trivializing sequences of completions (2.8) which yields the identity on Γ_i for any $i = 1, \ldots, n$, see the proof of Proposition 8.0.3(c). By Corollary 9.6.2 the associated birational transformation ψ of $B \times \mathbb{A}^1$ over B fitting in (9.26) is biregular. Due to our assumption $\mathcal{O}_B(B)^\times = \Bbbk^*$. Hence ψ is of the form

$$\psi: (b, u) \mapsto (b, \alpha u + \beta) \quad \text{where} \quad \alpha \in \Bbbk^* \quad \text{and} \quad \beta \in \mathcal{O}_B(B). \tag{10.10}$$

In particular, $\psi_*(z) = z$. Using z as a local parameter in B near b_i one can write

$$\beta(z) = \beta_{i,0} + \beta_{i,1} z + \ldots + \beta_{i,l} z^l + \ldots \quad \text{with} \quad \beta_{i,j} \in \Bbbk \ \forall i, j. \tag{10.11}$$

Consider the standard local charts $U(v) \subset X_l$ about $F(v)$ with local coordinates (z, t_v) and $U(w) \subset X_{l+1}$ about $F(w)$ with local coordinates (z, t_w) and, respectively, $U'(v) \subset X_l'$ about $F'(v)$ with local coordinates (z, t_v') and $U'(w) \subset X_{l+1}'$ about $F'(w)$ with local coordinates (z, t_w'). We claim that the restriction $\varphi|_{U(v)}: U(v) \mapsto U'(v)$ is given by

$$(z, t_v') = \psi_*(z, t_v) = (z, \alpha t_v + \beta_{i,l} + \beta_{i,l+1} z + \ldots). \tag{10.12}$$

Indeed, for $l = 0$, (10.12) follows from (10.10) and (10.11). Suppose by induction that (10.12) holds for a given $l \le h_i - 2$. Let $v, w \in \operatorname{vert}(\Gamma_i^*)$ with $l(v) = l$ and $l(w) = l + 1$ be joint by an edge, and let $c(w) = \varrho_{l+1}(F(w)) \in F(v)$. Applying the inductive hypothesis one obtains

$$\varphi_*: (z, t_v) \mapsto (z, \alpha t_v + \beta_{i,l} + \beta_{i,l+1} z + \beta_{i,l+2} z^2 + \ldots), \ c(w) \mapsto \alpha c(w) + \beta_{i,l}. \tag{10.13}$$

Using (10.3) and (10.13) one gets

$$\begin{aligned}\varphi(z, t_w) = (z, (t_v - c(w))/z) &\xrightarrow{\varphi_*} (z, \alpha(t_v - c(w))/z + \beta_{i,l+1} + \beta_{i,l+2} z + \ldots) \\ &= (z, \alpha t_w + \beta_{i,l+1} + \beta_{i,l+2} z + \ldots).\end{aligned}$$

This gives the inductive step. Letting $z = 0$ in (10.13) yields (10.9). □

PROPOSITION 10.4.5. (a) *The action of the group* $G(\mathcal{D})$ *on* $\mathcal{S}(\mathcal{D})$ *defined in* 10.3.3 *admits a lift to an action of* $G(\mathcal{D})$ *on* $\mathfrak{X}_l(\mathcal{D})$ *over* B *making the following morphisms in* (10.6) $G(\mathcal{D})$*-equivariant:*

$$\mathfrak{X}_l(\mathcal{D}) \to \mathcal{S}(\mathcal{D}),\ l = 0,\dots,m \quad \text{and} \quad r_l\colon \mathfrak{X}_l(\mathcal{D}) \to \mathfrak{X}_{l-1}(\mathcal{D}),\ l = 1,\dots,m. \tag{10.14}$$

(b) *Letting* $\mathfrak{X}(\mathcal{D}) = \mathfrak{X}_m(\mathcal{D})$ *one has* $\mathrm{Aut}_B(\mathfrak{X}(\mathcal{D}) \to \mathcal{S}(\mathcal{D})) = G(\mathcal{D})$.

(c) *For* $\mathfrak{s}, \mathfrak{s}' \in \mathcal{S}(\mathcal{D})$ *the surfaces* $X(\mathfrak{s})$ *and* $X(\mathfrak{s}')$ *are isomorphic over* B *if an only if* $\mathfrak{s}$ *and* $\mathfrak{s}'$ *lie on the same* $G(\mathcal{D})$*-orbit.*

(d) *The retraction* $\eta\colon \mathcal{S}(\mathcal{D}) \to \mathcal{S}^\circ(\mathcal{D}) \cong \mathcal{S}(\mathcal{D})/\mathbb{G}_m^{h(\mathcal{D})}$, *see Definition* 10.3.3, *defines a trivial* $\mathbb{G}_m^{h(\mathcal{D})}$*-bundle over* $\mathcal{S}^\circ(\mathcal{D})$.

(e) *The family* $\mathfrak{F}(\mathcal{D}) = (\mathfrak{X}(\mathcal{D}) \to \mathcal{S}(\mathcal{D}))$ *is isomorphic to the family induced from the restriction* $\mathfrak{F}^\circ(\mathcal{D}) = (\mathfrak{X}^\circ(\mathcal{D}) \to \mathcal{S}^\circ(\mathcal{D}))$ *via the retraction morphism* η. *Consequently, the deformation family* $\mathfrak{F}^\circ(\mathcal{D})$ *of GDF surfaces of type* $(B, z, \mathcal{D})$ *is versal with respect to the étale topology.*

PROOF. (a) We proceed by induction on l. The assertion is trivially true for $l = 0$. Suppose it holds for some $l \in \{0,\dots,m-1\}$. Then $g \in G(\mathcal{D})$ defines for any $\mathfrak{s} \in \mathcal{S}(\mathcal{D})$ an isomorphism

$$g|_{X_l(\mathfrak{s})}\colon X_l(\mathfrak{s}) \xrightarrow{\cong_B} X_l(g(\mathfrak{s}))$$

satisfying (10.9), see Lemma 10.4.4. The component $\mathfrak{s}_l \in \mathcal{S}_l(\mathcal{D})$ of $\mathfrak{s}$ is sent to the component $g(\mathfrak{s})_l \in \mathcal{S}_l(\mathcal{D})$ of $g(\mathfrak{s})$. It follows that the center $\mathfrak{C}_l(\mathcal{D}) \subset \mathfrak{X}_l(\mathcal{D})$ of the affine modification $r_l(\mathcal{D})\colon \mathfrak{X}_{l+1}(\mathcal{D}) \to \mathfrak{X}_l(\mathcal{D})$ is $G(\mathcal{D})$-invariant (cf. 10.4.2). Since its divisor $z^*(0)$ is $G(\mathcal{D})$-invariant too, due to Lemma 1.4.3 the action of $G(\mathcal{D})$ on $\mathfrak{X}_l(\mathcal{D})$ admits a lift to $\mathfrak{X}_{l+1}(\mathcal{D})$ making the morphisms (10.14) $G(\mathcal{D})$-equivariant.

(b) By (a) one has $\mathrm{Aut}_B(\mathfrak{X}(\mathcal{D}) \to \mathcal{S}(\mathcal{D})) \supset G(\mathcal{D})$. Applying the same argument to $g \in \mathrm{Aut}_B(\mathfrak{X}(\mathcal{D}) \to \mathcal{S}(\mathcal{D}))$ one concludes that g satisfies (10.9), hence belongs to $G(\mathcal{D})$.

(c) By virtue of (a), if $\mathfrak{s}$ and $\mathfrak{s}'$ lie on the same $G(\mathcal{D})$-orbit then $X(\mathfrak{s}) \cong_B X(\mathfrak{s}')$. Suppose further that $X(\mathfrak{s}) \cong_B X(\mathfrak{s}')$. Composing this isomorphism, say, φ with a suitable $\alpha \in \mathrm{Aut}_B(\mathcal{D})$ acting on $\mathfrak{X}(\mathcal{D})$ one may assume that φ induces the identity on $\mathcal{D}$. Then by Lemma 10.4.4, φ satisfies (10.9), and so, extends to an element of $G(\mathcal{D})$ acting on $\mathfrak{X}(\mathcal{D})$. Since $\mathfrak{X}(\mathcal{D}) \to \mathcal{S}(\mathcal{D})$ is $G(\mathcal{D})$-equivariant it follows that $\mathfrak{s}$ and $\mathfrak{s}'$ lie on the same $G(\mathcal{D})$-orbit. □

(d) The $\mathbb{G}_a^{h(\mathcal{D})}$-equivariant isomorphism

$$\Phi\colon \mathcal{S}^\circ(\mathcal{D}) \times \mathbb{G}_a^{h(\mathcal{D})} \xrightarrow{\cong} \mathcal{S}(\mathcal{D}), \quad (\mathfrak{s}, g) \mapsto g(\mathfrak{s})$$

yields the desired equivariant trivialization.

(e) Let $\eta^*(\mathfrak{F}^\circ(\mathcal{D})) = (\mathfrak{X}'(\mathcal{D})) \to \mathcal{S}(\mathcal{D}))$ be the induced family. Since $\mathcal{S}^\circ(\mathcal{D})$ is a slice for the $\mathbb{G}_a^{h(\mathcal{D})}$-action on $\mathcal{S}(\mathcal{D})$ and the projection $\pi_{\mathfrak{X}(\mathcal{D})}\colon \mathfrak{X}(\mathcal{D}) \to \mathcal{S}(\mathcal{D})$ is $\mathbb{G}_a^{h(\mathcal{D})}$-equivariant, see (a), then also $\mathfrak{X}^\circ(\mathcal{D})$ is a slice for the free $\mathbb{G}_a^{h(\mathcal{D})}$-action on $\mathfrak{X}(\mathcal{D})$, cf. (b). Therefore, one has a commutative diagram of $\mathbb{G}_a^{h(\mathcal{D})}$-equivariant morphisms

$$\begin{array}{ccc} \mathfrak{X}(\mathcal{D})^{\circ} \times \mathbb{G}_a^{h(\mathcal{D})} & \overset{\widetilde{\Phi}}{\underset{\cong}{\longrightarrow}} & \mathfrak{X}(\mathcal{D}) \\ \downarrow & & \downarrow \\ \mathcal{S}^{\circ}(\mathcal{D}) \times \mathbb{G}_a^{h(\mathcal{D})} & \overset{\cong}{\underset{\Phi}{\longrightarrow}} & \mathcal{S}(\mathcal{D}) \end{array} \tag{10.15}$$

where $\widetilde{\Phi}: (x, g) \mapsto g(x)$. Using Proposition 10.4.3 and the latter diagram the assertions follow.

REMARKS 10.4.6. 1. The action of the unipotent radical $R_u(G(\mathcal{D})) = \mathbb{G}_a^{h(\mathcal{D})}$ on $\mathfrak{X}(\mathcal{D})$ has the following interpretation. The group $\mathrm{Aut}_B(B \times \mathbb{A}^1) = \mathbb{U}_0 \rtimes \mathbb{G}_m$ where $\mathbb{U}_0 = \mathcal{O}_B(B)$ acts naturally on $\mathcal{S}(\mathcal{D})$ via (10.10). For $\mathfrak{s} \in \mathcal{S}(\mathcal{D})$ the subgroup $\mathbb{U}_{X(\mathfrak{s})} = \mathbb{U}_m$ from Proposition 10.2.2 is the stabilizer of $\mathfrak{s}$ in $\mathbb{U}$. By Proposition 10.2.2(c) for any $\mathfrak{s}' \in \mathcal{S}(\mathcal{D})$ one has

$$\mathbb{U}_{X(\mathfrak{s}')} = \mathbb{U}_{X(\mathfrak{s})} \cong H^0(B, -D_m) \quad \text{where} \quad D_m = \sum_{i=1}^{n} h_i b_i \,.$$

By Proposition 10.2.2(d), $\mathbb{U}_X = \mathbb{U}_{X(\mathfrak{s})}$ acts identically on $\mathcal{S}(\mathcal{D})$. The quotient $\mathbb{U}_0/\mathbb{U}_X \cong \mathbb{G}_a^{h(\mathcal{D})}$ acts transitively on each orbit of $\mathbb{U}_0$ in $\mathcal{S}(\mathcal{D})$, and this action coincides with the $\mathbb{G}_a^{h(\mathcal{D})}$-action.

2. It can be shown that the quotient $\mathfrak{M}(\mathcal{D}) = \mathcal{S}(\mathcal{D})/G(\mathcal{D})$ does not depend on the choice of a trivializing sequence for $\pi_X: X \to B$. Anyway, this fact follows also from the proof of Theorem 10.1.3(a) below.

10.5. Proof of Theorem 10.1.3

(a) We claim that the desired coarse moduli space is

$$\mathcal{C}(B, z, \mathcal{D}) = \mathfrak{M}(\mathcal{D}) = \mathcal{S}(\mathcal{D})/G(\mathcal{D}) = \mathcal{S}^{\circ}(\mathcal{D})/(\mathbb{G}_m \times \mathrm{Aut}_B^*(\mathcal{D}))\,,$$

see (10.5). Indeed, consider a deformation family $\mathfrak{F} : (\eta: \mathfrak{X} \to \mathcal{S},\ \pi: \mathfrak{X} \to B)$ of marked GDF surfaces $\pi|_{X(\mathfrak{s})}: X(\mathfrak{s}) = \eta^{-1}(\mathfrak{s}) \to B$ sharing the common marking $z \in \mathcal{O}_B(B)$ and the common graph divisor $\mathcal{D}(\pi|_{X(\mathfrak{s})}) \cong_B \mathcal{D}$ where the latter isomorphism depends on $\mathfrak{s} \in \mathcal{S}$.

Passing to the Galois covering $\mathcal{S}' \to \mathcal{S}$ defined by the monodromy group of the latter family we obtain the induced family $\mathfrak{F}' : (\mathfrak{X}' \to \mathcal{S}')$ with a trivial family of graph divisors. By Proposition 10.4.3 the configuration invariant defines a morphism $\Delta: \mathcal{S}' \to \mathcal{S}(\mathcal{D})$ such that $\mathfrak{F}' = \Delta^*(\mathfrak{F}(\mathcal{D}))$. The quotient morphism $\mathcal{S}' \to \mathcal{S}(\mathcal{D})/G(\mathcal{D})$ is constant on any fiber of $\mathcal{S}' \to \mathcal{S}$. Hence Δ can be factorized via a morphism $\delta: \mathcal{S} \to \mathcal{S}(\mathcal{D})/G(\mathcal{D}) = \mathfrak{M}(\mathcal{D})$. Due to Proposition 10.4.5(c) such a morphism δ is uniquely defined and has the desired properties, see Definition 10.1.2.

Statement (b) follows from Theorem 0.0.6.

Statement (c) is immediate by Proposition 7.2.2. Indeed, by virtue of this proposition, performing a top level stretching one replaces $\mathcal{D}$ by $\mathcal{D}^{(k)}$ without changing the isomorphism class over B of the cylinder $X \times \mathbb{A}^1$. Since X has a reducible fiber the graph divisor $\mathcal{D}^{(k)}$ of the resulting marked GDF surface $X^{(k)} \to B$ with the same marking z has at least $2k$ additional edges. So, one has

$$\dim \mathfrak{M}(\mathcal{D}) \geq \mathrm{card}\,(\mathrm{edges}\,(\mathcal{D})) - h(\mathcal{D}) - 1 \quad \text{and} \quad \dim \mathfrak{M}(\mathcal{D}^{(k)}) \geq \dim \mathfrak{M}(\mathcal{D}) + k\,,$$

which implies the assertion. □

COROLLARY 10.5.1. *Any deformation family of GDF surfaces whose graph divisors are chain divisors is trivial.*

PROOF. Let a GDF surface $\pi_X: X \to B$ admits a line bundle structure, that is, $\mathcal{D}(\pi_X) = \mathcal{D}$ is a chain divisor. Then $\operatorname{card}(\operatorname{edges}(\mathcal{D})) = h(\mathcal{D})$, $\mathcal{S}(\mathcal{D}) \cong \mathbb{A}^{h(\mathcal{D})}$, and the $\mathbb{G}_a^{h(\mathcal{D})}$-action on $\mathcal{S}(\mathcal{D})$ is simply transitive. Hence $\mathcal{S}^\circ(\mathcal{D})$ is a singleton, and so, the versal deformation family $\mathfrak{X}(\mathcal{D}) \to \mathcal{S}(\mathcal{D})$ is trivial by Proposition 10.4.5(d). Furthermore, any locally trivial family of chain divisors is trivial since the group $\operatorname{Aut}(\mathcal{D})$ is. Since any deformation family $\mathfrak{F} : \mathfrak{X} \to \mathcal{S}$ with the given graph divisor $\mathcal{D}$ is induced from $\mathfrak{X}(\mathcal{D}) \to \mathcal{S}(\mathcal{D})$ via the morphism $\Delta : \mathcal{S} \to \mathcal{S}(\mathcal{D})$ given by the configuration invariant, see Proposition 10.4.3(e), $\mathfrak{F}$ is trivial as well. □

REMARK 10.5.2. In fact, this corollary holds without any assumption on B. Theorem 10.1.3 remains valid if one replaces the assumption "$\mathcal{O}_B(B)^\times = \mathbb{k}^*$" by the following one: "$z^{-1}(0)$ is a singleton". However, in general the coarse moduli space $\mathcal{C}_1(B, z, \mathcal{D})$ does not exist in any reasonable category of spaces. Let us give a simple example.

EXAMPLE 10.5.3. Let $\mathbb{k} = \mathbb{C}$, and let $B = \mathbb{A}^1 \smallsetminus \{0, \pi\}$, where $\mathbb{A}^1 = \operatorname{Spec} \mathbb{C}[t]$, be equipped with the marking $z = t^2 - 1$ so that $n = 2$ and $b_1 = 1$, $b_2 = -1$. Let also Γ_i, $i = 1, 2$, be the rooted tree of height 1 with two vertices on level 1. One has

$$\mathcal{S}(\Gamma_i) = \mathcal{S}(2), \quad \mathcal{S}^\circ(\Gamma_i) \cong \mathbb{A}^1_* := \mathbb{A}^1 \smallsetminus \{0\},\ i = 1, 2, \quad \text{and} \quad \mathcal{S}^\circ(\mathcal{D}) \cong (\mathbb{A}^1_*)^2 .$$

The group $\operatorname{Aut}^*(\mathcal{D})$ is trivial, see Definition 10.3.3. The infinite discrete group

$$\mathcal{O}_B(B)^\times / \mathbb{k}^* \cong \mathbb{Z}^2$$

acts naturally on the quotient

$$\mathcal{S}^\circ(\mathcal{D}) / (\mathbb{G}_m \times \operatorname{Aut}^*(\mathcal{D})) \cong (\mathbb{A}^1_*)^2 / \mathbb{G}_m \cong \mathbb{A}^1_*$$

as a subgroup of $\mathbb{G}_m \subset \operatorname{Aut}(\mathbb{A}^1_*)$ generated by two nonzero complex numbers whose ratio is transcendental. Its orbits correspond to the isomorphism classes of the associated GDF surfaces. It is easily seen that the induced complex topology of the quotient $\mathbb{A}^1_* / \mathbb{Z}^2$ does not satisfy the Kolmogorov T_0 axiom. Hence this quotient neither is an algebraic (or complex) space, nor is an algebraic stack.

Acknowledgments

We are grateful to the anonymous referee for valuable remarks and suggestions. This work was partially done during our stays at the Max Planck Institute of Mathematics in Bonn in 2014–2015. We thanks this Institute for its generosity and excellent working conditions.

Bibliography

[1] S. S. Abhyankar, W. Heinzer, and P. Eakin, *On the uniqueness of the coefficient ring in a polynomial ring*, J. Algebra **23** (1972), 310–342, DOI 10.1016/0021-8693(72)90134-2. MR306173

[2] I. Arzhantsev, H. Flenner, S. Kaliman, F. Kutzschebauch, and M. Zaidenberg, *Flexible varieties and automorphism groups*, Duke Math. J. **162** (2013), no. 4, 767–823, DOI 10.1215/00127094-2080132. MR3039680

[3] I. Arzhantsev, U. Derenthal, J. Hausen, and A. Laface, *Cox rings*, Cambridge Studies in Advanced Mathematics, vol. 144, Cambridge University Press, Cambridge, 2015. MR3307753

[4] I. V. Arzhantsev and S. A. Gaĭfullin, *Cox rings, semigroups, and automorphisms of affine varieties* (Russian, with Russian summary), Mat. Sb. **201** (2010), no. 1, 3–24, DOI 10.1070/SM2010v201n01ABEH004063; English transl., Sb. Math. **201** (2010), no. 1-2, 1–21. MR2641086

[5] I. V. Arzhantsev, M. G. Zaĭdenberg, and K. G. Kuyumzhiyan, *Flag varieties, toric varieties, and suspensions: three examples of infinite transitivity* (Russian, with Russian summary), Mat. Sb. **203** (2012), no. 7, 3–30, DOI 10.1070/SM2012v203n07ABEH004248; English transl., Sb. Math. **203** (2012), no. 7-8, 923–949. MR2986429

[6] T. Asanuma, *Non-linearizable algebraic k^*-actions on affine spaces*, Invent. Math. **138** (1999), no. 2, 281–306, DOI 10.1007/s002220050379. MR1720185

[7] T. M. Bandman and L. Makar-Limanov, *Cylinders over affine surfaces*, Japan. J. Math. (N.S.) **26** (2000), no. 1, 207–217, DOI 10.4099/math1924.26.207. MR1771430

[8] T. Bandman and L. Makar-Limanov, *Affine surfaces with* $\mathrm{AK}(S) = \mathbb{C}$, Michigan Math. J. **49** (2001), no. 3, 567–582, DOI 10.1307/mmj/1012409971. MR1872757

[9] T. Bandman and L. Makar-Limanov, *Nonstability of the AK invariant*, Michigan Math. J. **53** (2005), no. 2, 263–281, DOI 10.1307/mmj/1123090767. MR2152699

[10] T. Bandman, L. Makar-Limanov. *Affine surfaces with isomorphic cylinders.* Unpublished notes, Bar Ilan University 2006, 17p.

[11] T. Bandman, L. Makar-Limanov. *Non-stability of AK-invariant for some* $\mathbb{Q}$*-planes.* Unpublished notes, Bar Ilan University 2006, 8p.

[12] A. Crachiola and L. Makar-Limanov, *On the rigidity of small domains*, J. Algebra **284** (2005), no. 1, 1–12, DOI 10.1016/j.jalgebra.2004.09.015. MR2115001

[13] A. J. Crachiola and L. G. Makar-Limanov, *An algebraic proof of a cancellation theorem for surfaces*, J. Algebra **320** (2008), no. 8, 3113–3119, DOI 10.1016/j.jalgebra.2008.03.037. MR2450715

[14] A. J. Crachiola and S. Maubach, *Rigid rings and Makar-Limanov techniques*, Comm. Algebra **41** (2013), no. 11, 4248–4266, DOI 10.1080/00927872.2012.695832. MR3169516

[15] D. A. Cox, *The homogeneous coordinate ring of a toric variety*, J. Algebraic Geom. **4** (1995), no. 1, 17–50. MR1299003

[16] D. Daigle, *Locally nilpotent derivations and Danielewski surfaces*, Osaka J. Math. **41** (2004), no. 1, 37–80. MR2040063

[17] W. Danielewski. *On a cancellation problem and automorphism groups of affine algebraic varieties.* Preprint, Warsaw, 1989.

[18] R. Dryło, *Non-uniruledness and the cancellation problem. II*, Ann. Polon. Math. **92** (2007), no. 1, 41–48, DOI 10.4064/ap92-1-4. MR2318509

[19] A. Dubouloz, *Danielewski-Fieseler surfaces*, Transform. Groups **10** (2005), no. 2, 139–162, DOI 10.1007/s00031-005-1004-x. MR2195597

[20] A. Dubouloz, *Embeddings of Danielewski surfaces in affine spaces*, Comment. Math. Helv. **81** (2006), no. 1, 49–73, DOI 10.4171/CMH/42. MR2208797

[21] A. Dubouloz. *Quelques remarques sur la notion de modification affine.* arXiv:math/0503142 (2005), 5p.

[22] A. Dubouloz, *The cylinder over the Koras-Russell cubic threefold has a trivial Makar-Limanov invariant*, Transform. Groups **14** (2009), no. 3, 531–539, DOI 10.1007/s00031-009-9051-3. MR2534798

[23] A. Dubouloz, *Flexible bundles over rigid affine surfaces*, Comment. Math. Helv. **90** (2015), no. 1, 121–137, DOI 10.4171/CMH/348. MR3317335

[24] A. Dubouloz, *Affine surfaces with isomorphic $\mathbb{A}^2$-cylinders*, Kyoto J. Math. **59** (2019), no. 1, 181–193, DOI 10.1215/21562261-2018-0005. MR3934627

[25] A. Dubouloz and P.-M. Poloni, *On a class of Danielewski surfaces in affine 3-space*, J. Algebra **321** (2009), no. 7, 1797–1812, DOI 10.1016/j.jalgebra.2008.12.009. MR2494748

[26] K.-H. Fieseler, *On complex affine surfaces with $\mathbf{C}^+$-action*, Comment. Math. Helv. **69** (1994), no. 1, 5–27, DOI 10.1007/BF02564471. MR1259603

[27] D. R. Finston and S. Maubach, *The automorphism group of certain factorial threefolds and a cancellation problem*, Israel J. Math. **163** (2008), 369–381, DOI 10.1007/s11856-008-0016-3. MR2391136

[28] H. Flenner and M. Zaidenberg, *Normal affine surfaces with $\mathbb{C}^*$-actions*, Osaka J. Math. **40** (2003), no. 4, 981–1009. MR2020670

[29] H. Flenner and M. Zaidenberg, *Locally nilpotent derivations on affine surfaces with a $\mathbb{C}^*$-action*, Osaka J. Math. **42** (2005), no. 4, 931–974. MR2196000

[30] H. Flenner, S. Kaliman, and M. Zaidenberg, *Completions of $\mathbb{C}^*$-surfaces*, Affine algebraic geometry, Osaka Univ. Press, Osaka, 2007, pp. 149–201. MR2327238

[31] H. Flenner, S. Kaliman, and M. Zaidenberg, *Uniqueness of $\mathbb{C}^*$- and $\mathbb{C}_+$-actions on Gizatullin surfaces*, Transform. Groups **13** (2008), no. 2, 305–354, DOI 10.1007/s00031-008-9014-0. MR2426134

[32] H. Flenner, S. Kaliman, and M. Zaidenberg, *Smooth affine surfaces with non-unique $\mathbb{C}^*$-actions*, J. Algebraic Geom. **20** (2011), no. 2, 329–398, DOI 10.1090/S1056-3911-2010-00533-4. MR2762994

[33] H. Flenner, S. Kaliman, and M. Zaidenberg. *Deformation equivalence of affine ruled surfaces.* arXiv:1305.5366v1 (2013), 34p.

[34] R. M. Fossum, *The divisor class group of a Krull domain*, Ergebnisse der Mathematik und ihrer Grenzgebiete, Band 74, Springer-Verlag, New York-Heidelberg, 1973. MR0382254

[35] T. Fujita, *On Zariski problem*, Proc. Japan Acad. Ser. A Math. Sci. **55** (1979), no. 3, 106–110. MR531454

[36] T. Fujita, *On the topology of noncomplete algebraic surfaces*, J. Fac. Sci. Univ. Tokyo Sect. IA Math. **29** (1982), no. 3, 503–566. MR687591

[37] J.-P. Furter and S. Lamy, *Normal subgroup generated by a plane polynomial automorphism*, Transform. Groups **15** (2010), no. 3, 577–610, DOI 10.1007/s00031-010-9095-4. MR2718938

[38] M. Furushima, *Finite groups of polynomial automorphisms in $\mathbf{C}^n$*, Tohoku Math. J. (2) **35** (1983), no. 3, 415–424, DOI 10.2748/tmj/1178229000. MR711357

[39] M. H. Gizatullin, *Quasihomogeneous affine surfaces* (Russian), Izv. Akad. Nauk SSSR Ser. Mat. **35** (1971), 1047–1071. MR0286791

[40] A. Grothendieck, *Fondements de la géométrie algébrique. [Extraits du Séminaire Bourbaki, 1957–1962.]* (French), Secrétariat mathématique, Paris, 1962. MR0146040

[41] R. V. Gurjar, K. Masuda, M. Miyanishi, and P. Russell, *Affine lines on affine surfaces and the Makar-Limanov invariant*, Canad. J. Math. **60** (2008), no. 1, 109–139, DOI 10.4153/CJM-2008-005-8. MR2381169

[42] R. V. Gurjar and M. Miyanishi, *Automorphisms of affine surfaces with $\mathbb{A}^1$-fibrations*, Michigan Math. J. **53** (2005), no. 1, 33–55, DOI 10.1307/mmj/1114021083. MR2125532

[43] R. Hartshorne, *Algebraic geometry*, Graduate Texts in Mathematics, No. 52, Springer-Verlag, New York-Heidelberg, 1977. MR0463157

[44] S. Iitaka and T. Fujita, *Cancellation theorem for algebraic varieties*, J. Fac. Sci. Univ. Tokyo Sect. IA Math. **24** (1977), no. 1, 123–127. MR450278

[45] S. Kaliman, *Actions of $\mathbb{C}^*$ and $\mathbb{C}_+$ on affine algebraic varieties*, Algebraic geometry—Seattle 2005. Part 2, Proc. Sympos. Pure Math., vol. 80, Amer. Math. Soc., Providence, RI, 2009, pp. 629–654, DOI 10.1090/pspum/080.2/2483949. MR2483949

[46] Sh. Kaliman and F. Kutzschebauch, *On algebraic volume density property*, Transform. Groups **21** (2016), no. 2, 451–478, DOI 10.1007/s00031-015-9360-7. MR3492044

[47] Sh. Kaliman and M. Zaidenberg, *Affine modifications and affine hypersurfaces with a very transitive automorphism group*, Transform. Groups **4** (1999), no. 1, 53–95, DOI 10.1007/BF01236662. MR1669174

[48] S. Kovalenko, A. Perepechko, and M. Zaidenberg, *On automorphism groups of affine surfaces*, Algebraic varieties and automorphism groups, Adv. Stud. Pure Math., vol. 75, Math. Soc. Japan, Tokyo, 2017, pp. 207–286, DOI 10.2969/aspm/07510207. MR3793368

[49] H. Lange, *On elementary transformations of ruled surfaces*, J. Reine Angew. Math. **346** (1984), 32–35, DOI 10.1515/crll.1984.346.32. MR727394

[50] V. Lin, M. Zaidenberg. *Automorphism groups of configuration spaces and discriminant varieties*. `arXiv:1505.06927` (2015), 61p.

[51] J. Lipman, *Rational singularities, with applications to algebraic surfaces and unique factorization*, Inst. Hautes Études Sci. Publ. Math. **36** (1969), 195–279. MR276239

[52] L. Makar-Limanov, *On the group of automorphisms of a surface* $x^n y = P(z)$, Israel J. Math. **121** (2001), 113–123, DOI 10.1007/BF02802499. MR1818396

[53] K. Masuda, *Families of hypersurfaces with noncancellation property*, Proc. Amer. Math. Soc. **145** (2017), no. 4, 1439–1452, DOI 10.1090/proc/13489. MR3601537

[54] K. Masuda and M. Miyanishi, *Affine pseudo-planes and cancellation problem*, Trans. Amer. Math. Soc. **357** (2005), no. 12, 4867–4883, DOI 10.1090/S0002-9947-05-04046-8. MR2165391

[55] K. Masuda and M. Miyanishi, *Equivariant cancellation for algebraic varieties*, Affine algebraic geometry, Contemp. Math., vol. 369, Amer. Math. Soc., Providence, RI, 2005, pp. 183–195, DOI 10.1090/conm/369/06812. MR2126662

[56] M. Miyanishi, *Open algebraic surfaces*, CRM Monograph Series, vol. 12, American Mathematical Society, Providence, RI, 2001, DOI 10.1090/crmm/012. MR1800276

[57] M. Miyanishi and T. Sugie, *Affine surfaces containing cylinderlike open sets*, J. Math. Kyoto Univ. **20** (1980), no. 1, 11–42, DOI 10.1215/kjm/1250522319. MR564667

[58] L. Moser-Jauslin and P.-M. Poloni, *Embeddings of a family of Danielewski hypersurfaces and certain* $\mathbf{C}^+$*-actions on* $\mathbf{C}^3$ (English, with English and French summaries), Ann. Inst. Fourier (Grenoble) **56** (2006), no. 5, 1567–1581. MR2273864

[59] D. Mumford, J. Fogarty, and F. Kirwan, *Geometric invariant theory*, 3rd ed., Ergebnisse der Mathematik und ihrer Grenzgebiete (2) [Results in Mathematics and Related Areas (2)], vol. 34, Springer-Verlag, Berlin, 1994, DOI 10.1007/978-3-642-57916-5. MR1304906

[60] M. Nagata and M. Maruyama, *Note on the structure of a ruled surface*, J. Reine Angew. Math. **239(240)** (1969), 68–73, DOI 10.1515/crll.1969.239-240.68. MR257085

[61] P.-M. Poloni, *Classification(s) of Danielewski hypersurfaces*, Transform. Groups **16** (2011), no. 2, 579–597, DOI 10.1007/s00031-011-9146-5. MR2806502

[62] V. L. Popov. *Open Problems*. In: Affine algebraic geometry, 12–16, Contemp. Math. 369, Amer. Math. Soc., Providence, RI, 2005.

[63] C. P. Ramanujam, *A note on automorphism groups of algebraic varieties*, Math. Ann. **156** (1964), 25–33, DOI 10.1007/BF01359978. MR166198

[64] W. Rudin, *Preservation of level sets by automorphisms of* $\mathbf{C}^n$, Indag. Math. (N.S.) **4** (1993), no. 4, 489–497, DOI 10.1016/0019-3577(93)90020-Y. MR1252994

[65] J.-P. Serre. *Espaces fibrés algébriques*. Séminaire C. Chevalley, Anneaux de Chow, Exposé 1, 1958.

[66] J.-P. Serre. *Sur les modules projectifs*. Sém. Dubreil-Pisot 14 (1960 – 61), 1–16.

[67] H. Sumihiro, *Equivariant completion*, J. Math. Kyoto Univ. **14** (1974), 1–28, DOI 10.1215/kjm/1250523277. MR337963

[68] T. tom Dieck, *Homology planes without cancellation property*, Arch. Math. (Basel) **59** (1992), no. 2, 105–114, DOI 10.1007/BF01190674. MR1170634

[69] J. Wilkens, *On the cancellation problem for surfaces* (English, with English and French summaries), C. R. Acad. Sci. Paris Sér. I Math. **326** (1998), no. 9, 1111–1116, DOI 10.1016/S0764-4442(98)80071-2. MR1647227

[70] D. Wright, *Polynomial automorphism groups*, Polynomial automorphisms and related topics, Publishing House for Science and Technology, Hanoi, 2007, pp. 1–19. MR2389269

[71] M. Zaidenberg. Lecture course *"Affine surfaces and the Zariski Cancellation Problem"* (a program). `http://www.mat.uniroma2.it/~flamini/workshops/LectZaidenberg.html`

[72] O. Zariski, *The reduction of the singularities of an algebraic surface*, Ann. of Math. (2) **40** (1939), 639–689, DOI 10.2307/1968949. MR159

Editorial Information

To be published in the *Memoirs*, a paper must be correct, new, nontrivial, and significant. Further, it must be well written and of interest to a substantial number of mathematicians. Piecemeal results, such as an inconclusive step toward an unproved major theorem or a minor variation on a known result, are in general not acceptable for publication.

Papers appearing in *Memoirs* are generally at least 80 and not more than 200 published pages in length. Papers less than 80 or more than 200 published pages require the approval of the Managing Editor of the Transactions/Memoirs Editorial Board. Published pages are the same size as those generated in the style files provided for $\mathcal{A}_{\mathcal{M}}\mathcal{S}$-LaTeX.

Information on the backlog for this journal can be found on the AMS website starting from `http://www.ams.org/memo`.

A Consent to Publish is required before we can begin processing your paper. After a paper is accepted for publication, the Providence office will send a Consent to Publish and Copyright Agreement to all authors of the paper. By submitting a paper to the *Memoirs*, authors certify that the results have not been submitted to nor are they under consideration for publication by another journal, conference proceedings, or similar publication.

Information for Authors

Memoirs is an author-prepared publication. Once formatted for print and on-line publication, articles will be published as is with the addition of AMS-prepared frontmatter and backmatter. Articles are not copyedited; however, confirmation copy will be sent to the authors.

Initial submission. The AMS uses Centralized Manuscript Processing for initial submissions. Authors should submit a PDF file using the Initial Manuscript Submission form found at `www.ams.org/submission/memo`, or send one copy of the manuscript to the following address: Centralized Manuscript Processing, MEMOIRS OF THE AMS, 201 Charles Street, Providence, RI 02904-2294 USA. If a paper copy is being forwarded to the AMS, indicate that it is for *Memoirs* and include the name of the corresponding author, contact information such as email address or mailing address, and the name of an appropriate Editor to review the paper (see the list of Editors below).

The paper must contain a *descriptive title* and an *abstract* that summarizes the article in language suitable for workers in the general field (algebra, analysis, etc.). The *descriptive title* should be short, but informative; useless or vague phrases such as "some remarks about" or "concerning" should be avoided. The *abstract* should be at least one complete sentence, and at most 300 words. Included with the footnotes to the paper should be the 2020 *Mathematics Subject Classification* representing the primary and secondary subjects of the article. The classifications are accessible from `www.ams.org/msc/`. The Mathematics Subject Classification footnote may be followed by a list of *key words and phrases* describing the subject matter of the article and taken from it. Journal abbreviations used in bibliographies are listed in the latest *Mathematical Reviews* annual index. The series abbreviations are also accessible from `www.ams.org/msnhtml/serials.pdf`. To help in preparing and verifying references, the AMS offers MR Lookup, a Reference Tool for Linking, at `www.ams.org/mrlookup/`.

Electronically prepared manuscripts. The AMS encourages electronically prepared manuscripts, with a strong preference for $\mathcal{A}_{\mathcal{M}}\mathcal{S}$-LaTeX. To this end, the Society has prepared $\mathcal{A}_{\mathcal{M}}\mathcal{S}$-LaTeX author packages for each AMS publication. Author packages include instructions for preparing electronic manuscripts, samples, and a style file that generates the particular design specifications of that publication series.

Authors may retrieve an author package for *Memoirs of the AMS* from `www.ams.org/journals/memo/memoauthorpac.html`. The *AMS Author Handbook* is available in PDF format from the author package link. The author package can also be obtained free of charge by sending email to `tech-support@ams.org` or from the Publication Division,

American Mathematical Society, 201 Charles St., Providence, RI 02904-2294, USA. When requesting an author package, please specify the publication in which your paper will appear. Please be sure to include your complete mailing address.

After acceptance. The source files for the final version of the electronic manuscript should be sent to the Providence office immediately after the paper has been accepted for publication. The author should also submit a PDF of the final version of the paper to the editor, who will forward a copy to the Providence office.

Accepted electronically prepared files can be submitted via the web at `www.ams.org/submit-book-journal/`, sent via FTP, or sent on CD to the Electronic Prepress Department, American Mathematical Society, 201 Charles Street, Providence, RI 02904-2294 USA. TEX source files and graphic files can be transferred over the Internet by FTP to the Internet node `ftp.ams.org` (130.44.1.100). When sending a manuscript electronically via CD, please be sure to include a message indicating that the paper is for the *Memoirs*.

Electronic graphics. Comprehensive instructions on preparing graphics are available at `www.ams.org/authors/journals.html`. A few of the major requirements are given here.

Submit files for graphics as EPS (Encapsulated PostScript) files. This includes graphics originated via a graphics application as well as scanned photographs or other computer-generated images. If this is not possible, TIFF files are acceptable as long as they can be opened in Adobe Photoshop or Illustrator.

Authors using graphics packages for the creation of electronic art should also avoid the use of any lines thinner than 0.5 points in width. Many graphics packages allow the user to specify a "hairline" for a very thin line. Hairlines often look acceptable when proofed on a typical laser printer. However, when produced on a high-resolution laser imagesetter, hairlines become nearly invisible and will be lost entirely in the final printing process.

Screens should be set to values between 15% and 85%. Screens which fall outside of this range are too light or too dark to print correctly. Variations of screens within a graphic should be no less than 10%.

Any graphics created in color will be rendered in grayscale for the printed version unless color printing is authorized by the Managing Editor and the Publisher. In general, color graphics will appear in color in the online version.

Inquiries. Any inquiries concerning a paper that has been accepted for publication should be sent to `memo-query@ams.org` or directly to the Electronic Prepress Department, American Mathematical Society, 201 Charles St., Providence, RI 02904-2294 USA.

Editors

This journal is designed particularly for long research papers, normally at least 80 pages in length, and groups of cognate papers in pure and applied mathematics. Papers intended for publication in the *Memoirs* should be addressed to one of the following editors. The AMS uses Centralized Manuscript Processing for initial submissions to AMS journals. Authors should follow instructions listed on the Initial Submission page found at www.ams.org/memo/memosubmit.html.

SELECTED PUBLISHED TITLES IN THIS SERIES

1365 **Alessandro Andretta and Luca Motto Ros,** Souslin Quasi-Orders and Bi-Embeddability of Uncountable Structures, 2022

1364 **Marco De Renzi,** Non-Semisimple Extended Topological Quantum Field Theories, 2022

1363 **Alan Hammond,** Brownian Regularity for the Airy Line Ensemble, and Multi-Polymer Watermelons in Brownian Last Passage Percolation, 2022

1362 **John Voight and David Zureick-Brown,** The Canonical Ring of a Stacky Curve, 2022

1361 **Nuno Freitas and Alain Kraus,** On the Symplectic Type of Isomorphisms of the p-Torsion of Elliptic Curves, 2022

1360 **Alexander V. Kolesnikov and Emanuel Milman,** Local L^p-Brunn-Minkowski Inequalities for $p < 1$, 2022

1359 **Franck Barthe and Paweł Wolff,** Positive Gaussian Kernels Also Have Gaussian Minimizers, 2022

1358 **Swee Hong Chan and Lionel Levine,** Abelian Networks IV. Dynamics of Nonhalting Networks, 2022

1357 **Camille Laurent and Matthieu Léautaud,** Tunneling Estimates and Approximate Controllability for Hypoelliptic Equations, 2022

1356 **Matthias Grüninger,** Cubic Action of a Rank One Group, 2022

1355 **David A. Craven,** Maximal PSL_2 Subgroups of Exceptional Groups of Lie Type, 2022

1354 **Gian Paolo Leonardi, Manuel Ritoré, and Efstratios Vernadakis,** Isoperimetric Inequalities in Unbounded Convex Bodies, 2022

1353 **Clifton Cunningham, Andrew Fiori, Ahmed Moussaoui, James Mracek, and Bin Xu,** Arthur Packets for p-adic Groups by Way of Microlocal Vanishing Cycles of Perverse Sheaves, with Examples, 2022

1352 **Bernhard Mühlherr, Richard Weiss, and Holger P. Petersson,** Tits Polygons, 2022

1351 **Athanassios S. Fokas and Jonatan Lenells,** On the Asymptotics to all Orders of the Riemann Zeta Function and of a Two-Parameter Generalization of the Riemann Zeta Function, 2022

1350 **Çağatay Kutluhan, Steven Sivek, and C. H. Taubes,** Sutured ECH is a Natural Invariant, 2022

1349 **Leonard Gross,** The Yang-Mills Heat Equation with Finite Action in Three Dimensions, 2022

1348 **Murat Akman, Jasun Gong, Jay Hineman, John Lewis, and Andrew Vogel,** The Brunn-Minkowski Inequality and a Minkowski Problem for Nonlinear Capacity, 2022

1347 **Zhiwu Lin and Chongchun Zeng,** Instability, Index Theorem, and Exponential Trichotomy for Linear Hamiltonian PDEs, 2022

1342 **Haosui Duanmu, Jeffrey S. Rosenthal, and William Weiss,** Ergodicity of Markov Processes via Nonstandard Analysis, 2018

1341 **Mima Stanojkovski,** Intense Automorphisms of Finite Groups, 2021

1340 **Le Chen, Yaozhong Hu, and David Nualart,** Regularity and Strict Positivity of Densities for the Nonlinear Stochastic Heat Equations, 2021

1339 **Gong Chen,** Strichartz Estimates for Wave Equations with Charge Transfer Hamiltonians, 2021

1338 **András Juhász, Dylan P. Thurston, and Ian Zemke,** Naturality and Mapping Class Groups in Heegard Floer Homology, 2021

1323 **Shu Kawaguchi and Kazuhiko Yamaki,** Effective Faithful Tropicalizations Associated to Linear Systems on Curves, 2021

For a complete list of titles in this series, visit the
AMS Bookstore at **www.ams.org/bookstore/memoseries/**.